Pharmaceutical Isolators

Pharmaceutical Isolators

A guide to their application, design and control

Edited by the Pharmaceutical
Isolator Working Party

A Working Party of the UK NHS Pharmaceutical
Quality Assurance Committee

Principal Editors for the Pharmaceutical
Isolator Working Party

Brian Midcalf

BPharm, FRPharmS
Assistant PTQA Course Director
School of Continuing Education, University of Leeds, UK

W Mitchell Phillips

BPharm, MRPharmS
West Midlands Quality Assurance Pharmacist
Birmingham, UK

John S Neiger

MA, CEng, MIMechE, MSEE
Chairman
Envair Limited, Haslingden, UK

Tim J Coles

BSc, MPhil
Isolator Specialist
GRC Consultants, Alton, UK

London • Chicago Pharmaceutical Press

Published by the Pharmaceutical Press
Publications division of the Royal Pharmaceutical Society of Great Britain

1 Lambeth High Street, London SE1 7JN, UK
100 South Atkinson Road, Suite 206, Grayslake, IL 60030-7820, USA

Text design by Barker/Hilsdon, Lyme Regis, Dorset
Typeset by Type Study, Scarborough, North Yorkshire
Printed in Great Britain by TJ International, Padstow, Cornwall

ISBN 0 85369 573 3

A catalogue record for this book is available from the British Library

Contents

Preface

It is a great privilege to be involved with this new guide for pharmaceutical isolators and it is hoped that its contents will be invaluable to designers, manufacturers, users and inspectors or auditors. The contents have been written and reviewed by members of the Pharmaceutical Isolator Working Party which includes experts from industry, hospitals and isolator manufacturers.

Some delay in preparation of the book occurred while the International Standards Organisation (ISO) was developing a style of nomenclature and a definition of terms that would be more universally acceptable. At the time of going to press, some of these terms have not been finally agreed but further delay in publication was unacceptable, since it could potentiate a perceived risk to patients or operators if they were adhering to older working practices.

On a historical note, the working party to review isolators was originally convened following discussions between UK NHS Regional Quality Control Pharmacists and some isolator manufacturers. The Regional Quality Control Pharmacists' subcommittee of the Regional Pharmaceutical Officers, later the Pharmaceutical Quality Control Committee of the NHS and now the NHS Quality Assurance Committee, agreed to the formation of a special interest group in 1992.

An open membership style was adopted with representation from interested and relevant bodies including quality control pharmacists, manufacturing pharmacists (hospital and industry), radiopharmacists, isolator manufacturers, Medicines and Healthcare products Regulatory Agency (MHRA) and consultants or advisers in the field of isolation technology. Liaison with any other bodies or specialist groups, both nationally or internationally, was encouraged and has developed over a number of years.

Approximately 40 meetings have taken place since the development of the group and the compilation of this new book. The Working Party agreed to the members forming specialist interest groups that have concentrated on many different topics. Care was taken to ensure that discussions to develop the paper were comprehensive and that a general agreement was reached which was acceptable to all members. The MHRA representation was used in a consultative capacity for the review of some of the text. We hope this has ensured that consistent and sound advice was maintained.

The overall objective was to consider the basic concepts, definitions and standards necessary in the design, construction, commissioning, maintenance and use of isolators. The book also considers as many associated activities as possible, to form a practical and useful guide.

This new book is a guide for intending purchasers, existing users, manufacturers and those involved in standard setting and monitoring. It focuses on isolators for pharmaceutical use. It is appreciated that other aspects of isolator design are important to particular groups of users, but the principles identified here will also generally be applicable to other specialist uses. The different types of isolator are described reflecting the deliberations of the CEN (European Committee for Standardization)/ISO Technical Committee and the more technical aspects of isolator design, commissioning and performance monitoring are detailed. The critical areas of passing items into and out of the isolator have been considered and the classification includes as many transfer systems as possible. The growing acceptance of DQ, IQ, OQ and PQ (and more recently RQ) terminology has been described (see Chapter 10).

Over recent years, considerable concern has been expressed over the safe use of isolators whilst handling hazardous materials. The question is frequently asked whether it is safe for operators to process these hazardous substances in a positive or negative pressure isolator. Will the safety of the operator be compromised in the former or the integrity of the process be compromised in the latter? A special investigative group was set up with the collaboration of the MHRA and the Health and Safety Executive (HSE) to assess these factors. The preliminary results of a study were reported at the 5th Isolator Conference in April 2000 and further elaborated at the 6th Isolator Conference in October 2001. A more extensive study will be required to reach verifiable conclusions but there was no evidence of an immediate risk to the operator from using either type of isolator.

Considerable development has taken place over the past five years in isolator design and construction. A welcome dialogue between the manufacturers and the users has taken place. This has led to the production of isolators with much better user acceptability, that are easier to clean, maintain and use, and to the development of accessories which help in these aspects.

It is likely that guidance given in this book may now be in excess of that which can be achieved in systems with older designs of isolators. This is inevitable and common to any system that has to comply with a current quality standard. Users of existing older isolators should bear in mind the principles used in this document and decide for which purpose their isolator is now suitable and whether some upgrading is appropriate.

Most of the factors to be considered in planning a new isolator unit for industry or healthcare should include positioning, size, type of throughput, environmental controls, commissioning and monitoring. Procedural aspects are left to the managers of the departments concerned in developing their operation.

Inspectors and auditors have shown greater interest in systems using isolators. This probably follows a greater in-depth knowledge of the technology. Users now are expected to provide more evidence from their validation studies that the systems are working to their design and operational qualification. It is hoped that the more technical aspects of this book will help the users to provide the data and inspectors to appreciate and understand it.

Comments on this book will be welcome and readers and users should send their comments to the Working Party Chairman who will act as coordinator for such discussion.

Brian Midcalf
W Mitchell Phillips
John Neiger
Tim Coles
February 2004

Acknowledgement

The chairman wishes to acknowledge the collective skills and expertise shown by all members of the Working Party in the compilation of this document. The membership list is appended. Particular recognition is made to those involved with the generation of the different sections of the book and the generous dedication of time from Working Party members in assessing the status and accuracy of its content. Readers cannot be expected to appreciate the tremendous amount of discussion undertaken in its production. The tolerance and forbearance of all those who contributed are recognised with gratitude.

Brian Midcalf
February 2004

About the editors

Brian Midcalf, BPharm FRPharmS, has spent most of his full time employment as a Regional Quality Assurance Pharmacist, with over 30 years of service with the NHS as a QA Pharmacist and Qualified Person with Leeds Teaching Hospitals NHS Trust, St James's University Hospital, Leeds.

He has held a number of appointments to NHS Committees including Committee on Dental and Surgical Materials, Committee on Safety of Medicines Subcommittee Chemistry and Pharmacy Standards, and British Pharmacopoeia Committee B. Brian has served as chairman of National Quality Control Committee and as secretary of the Surgical Dressings Working Party of QCC and was a founding member of the regional NHS COSHH Committee. He is a member of BSI/ISO LBI-30 committee.

As an original member of the Course Management Committee for Pharmacy Technology and Quality Assurance (PTQA), Brian is a module director for a number of PTQA modules and for the development of some new courses. He is Assistant PTQA Course Director.

In support of the use of barrier technology in hospital pharmacy, Brian became Convenor and Chairman of the UK Pharmaceutical Isolator Working Party, has chaired seven Pharmaceutical Isolator Conferences and was joint editor of the HMSO publication 'Isolators for Pharmaceutical Applications'. He is a leading editor for the development of this new book on isolators.

Brian is registered as a Qualified Person with the RPSGB and is a Fellow of the Royal Pharmaceutical Society of Great Britain.

John Neiger, MA CEng MIMechE MSEE, was a founding director of Envair Limited in 1972 and is currently Chairman of that company. Envair is a specialist company engaged in the provision of clean air and containment environments for medicine, research and industry and its products range from simple 'laminar flow' clean air work stations, microbiological safety cabinets and isolators through to turnkey cleanrooms and containment suites.

John graduated in Mechanical Engineering at Cambridge University in 1961. He is a Chartered Engineer, Member of the Institution of Mechanical Engineers and Member of the Society of Environmental Engineers. The early part of his career was with the aim of preparing him to take a part in his family's textile business which he joined in 1967. When that was sold in 1971, he was able to return to mainstream engineering and became part of the team that was responsible for the creation and development of Envair.

In 1986 Envair entered the field of isolators and, in collaboration with the Royal Hallamshire Hospital, Sheffield and Hope Hospital, Salford, developed some of the original hospital pharmacy isolators. Shortly after the UK Pharmaceutical Isolator Working Party was set up by the NHS Regional QC Pharmacists Committee in 1993, John became a member and has remained so ever since.

John sits on the BSI committee responsible for UK input into the ISO 14644 'Cleanrooms and associated controlled environments' series of standards and is one of the UK technical experts on the ISO Working Group that drafted the part concerned with Isolators, Part 7: Separative devices (clean air hoods, gloveboxes, isolators, minienvironments). He also sits on the BSI committee which is responsible for the UK contribution to EN 12469 'Biotechnology – Performance criteria for microbiological safety cabinets'. He is a committee member of the Contamination Control Group of the Society of Environmental Engineers.

John has written a number of articles and has had two papers published in the European Journal of Parenteral & Pharmaceutical Sciences. He has been a speaker at conferences and courses in various parts of the world including ISOPP (International Society of Oncology Pharmacy Practitioners) Prague in 2000 and, on several occasions, the annual Aseptic Dispensing Course run by the Pharmaceutical Society of Singapore.

Tim Coles, BSc MPhil, is an Engineering Specialist and has been working for GRC Consultants, a division of the Mott MacDonald Group, since 1998. His main interest has been in pharmaceutical isolation technology, particularly in the field of aseptic operations and gas phase sanitisation. This work has included the design of facilities, specification of the equipment and subsequent validation.

Tim holds BSc and MPhil degrees in Environmental Sciences from the University of East Anglia. He has previously worked for equipment manufacturers including La Calhene GB and his own company, Cambridge Isolation Technology Ltd. His book, 'Isolation Technology – A Practical Guide' was published by Interpharm Press Inc (now CRC Inc) in 1998 with a second edition due out in 2004. Tim writes and lectures regularly on the subject of isolation technology.

W Mitchell Phillips, BSc(Pharm) MRPharmS, has spent most of his full time employment as a Regional Quality Assurance Pharmacist with the NHS as a QA Pharmacist and Qualified Person with Birmingham Teaching Hospitals NHS Trust, in the former West Midlands Region.

He has served as chairman and secretary of National Quality Control Committee and has been involved in working groups that were responsible for numerous advisory documents including the well accepted Aseptic Preparations guidelines and earlier drafts of guidance for pharmaceutical isolators.

Mitchell is registered as a Qualified Person with the RPSGB and has been involved with the training and education of technical pharmacists. He has played a leading role in technical pharmacy auditing and assists fellow members of the QCC in maintaining an annual report of this activity.

As current secretary of the Pharmaceutical Isolator Working Party, Mitch has played a vital role in establishing the working groups that developed this book and has commanded considerable respect for his editing skills in the compilation of the numerous disparate contributions.

Contributors

The technical content of the book has been achieved by very substantial contributions from people in the group and also some significant help from people outside the Working Party. These include:

Membership of the working party

Ken Baker	Consultant, gloves and gauntlet specialist
Andrew Bill	MRHA (MCA) Medicines Inspectorate
Russell Brammah	Consultant, Member of ISO TC209
Caroline Coles	Pharminox
Tim Coles	GRC Consultants
James Drinkwater	Bioquell Pharma
Beverley Ellis	Radiopharmacist, North West NHS
Mike Foster	Bassaire
Brian Midcalf	Pharmacy Consultant, University of Leeds (ex NHS Quality Assurance)
David Morley	GRC Consultants
John Neiger	Envair
Robert Parkinson	Quality Assurance, Trent NHS
Mitchell Phillips	Quality Assurance, West Midlands NHS
Mike Pickerill	La Calhene
Karen Rossington	Shield Medicare
Peter White	NOVA Laboratories
Graham Wilson	Amercare
Liz Allanson	MRHA (MCA) Medicines Inspectorate (consulting adviser)

Other contributors and consultants

Gordon Farquharson	Bovis LendLease (HEPA filters)
Craig Hughes	PQS (ex PQAS NHS – microbiological monitoring)

Wander ter Kuile	McLeod Russell (HEPA penetration diagram)
Carla Martin	Programme Manager, BSI (standards)
Gerald McDonnell	Steris (decontamination)
David Michael	BSI (standards)
Murray Nicholson	Steris (decontamination)
Robert Pringle	PQS (ex PQAS NHS – physical monitoring)
Barbara Salvage	Shield Medicare (decontamination)
Tom Sexton	Semper Products (carbon filters)
Suzanne Stubbs	Shield Medicare (decontamination)
Frank Thomas	BSI (standards)
Roy Venkatesh	CIVAS (carbon filters and stainless steel)
Mike Vinson	Camfil Farr (HEPA filters)
David Watling	BioQuell (decontamination)
HEPA penetration diagram	McLeod Russell
Particle size chart	Anderson Sampling

How to use this guide

Developing the theme of what you can find in this guide and how you might like to use it, this introductory chapter is written in a non-technical style. It is intended to help you to understand how the book is organised and where to look for particular items or topics. The style should make this section easy to read and not put off those who are not familiar with isolators. Readers who already have some knowledge should be able to go straight to the technical sections of the book. The editors hope you will be successful in your endeavours.

We are pleased you have chosen to use our book. We hope the sequence of chapters is in a logical order so that it is possible for you to gain an ever greater understanding of isolators. You may now be looking for advice on How to Use This Guide. It is our intention that you will be able to refer to it to gain information and deeper knowledge about pharmaceutical isolators, how to handle them, where to put them and how to look after them. It is recommended that the guide is used in its entirety as it places in context the many considerations needed to devise and maintain a specialist unit which uses isolators.

What action may be required in your own unit, company or hospital as a result of using this guide will be controlled by a number of factors. You will appreciate that it is not possible to generate a guide that can be immediately adoptable in every single type of user situation. We have sought to ensure that we are compliant with current standards and guidelines, which we have listed and referred to. At the same time we have sought not to be too prescriptive. There is sufficient guidance for designers of new units to build something that will comply with current good practice. On the other hand, it is accepted that some users will be working in units of 'lower specification'. The 'non-ideal' unit is recognised and this guide should give direction as to where improvements are needed without necessarily indicating that a unit is of unacceptable quality.

Licensed manufacturing units are subject to regulatory control. If these are found to be non-compliant with current standards and guidelines, they will require immediate action to allow production to continue.

In health service pharmacies, when non-compliance is found, a judgement will need to be made regarding the risk to the patient of continuing to use the unit, as opposed to preparing the product on the ward, providing it from an alternative source or not providing it at all. This comment is not aimed to encourage apathy but rather to keep things in perspective. Where 'non-compliance' is detected, it is strongly recommended that you ensure an action plan is generated, usually in agreement with your auditor or inspector, and that a good business case is made to improve and rectify all deficiencies.

Thus the guide can be used both for improving existing units and for designing, constructing and commissioning good new ones. It goes without saying that all operational units will be supported by good standard operating procedures (SOPs), validation, and environmental monitoring. This guide identifies ideals for new units, which existing units should strive to achieve.

'How To Use This Guide' is presented in an informal conversational style to encourage those who are unfamiliar with isolators and isolator technology, to approach what might be very challenging in a more confident manner.

The first step for you, the reader, could be to look at Chapter 1: **Isolator applications**, as that will indicate the types of process for which an isolator can provide the best environment. It is possible that there is a new activity that has been or is about to be imposed on your operation. This may have prompted you to consider the use of an isolator as a possible solution. An assessment of the processes currently undertaken in your operation and also those anticipated, together with the safest way of achieving the required output, will have to be made. The safety implication referred to here is the need to achieve optimal safety for both the operator and the product.

Most of the activities that use isolators, known to the authors, have been identified. As there are so many of these, both newly developing and in regular use, giving a firm recommendation for your particular requirement is impractical. A table has been compiled which should be useful, either as an answer or as a starting point for further guidance.

Assuming that you have now decided to consider the use of an isolator for your activity, which is likely to require the safer aseptic manipulation of components, the next step would be to study the **Design** chapter. This should identify a number of options that are available as we describe the many basic configurations of isolator construction. This chapter also provides information on some key elements such as filters and different airflow patterns.

An isolator requires an appropriate means for the transfer of items into and out of its controlled workspace. We refer to these as **Transfer devices.** It will only take a moment's thought to realise that a sealed and sanitised controlled workspace will probably stay uncontaminated if it remains sealed. So, to be of any practical use, there must a way of introducing or removing raw materials, components and finished products, without breaking the barrier maintained by the isolator and without compromising its clean integrity. Transfer devices achieve this vital link between the controlled workspace and the outside environment with only minimal contamination of either.

Manufacturers have developed a number of different ways of constructing and configuring transfer devices. An original classification was made in 'Isolators for Pharmaceutical Applications, Edition 2' and this classification has been adopted in the new Draft ISO 14644-Part 7 standard, with only minor changes. It is important to know what devices are available, their characteristics and their limitations.

So far, you have an isolator which allows you to introduce and remove the materials you wish to manipulate. Now you need to be able to access the controlled workspace to carry out your manipulations. The **Access devices** chapter explains how you can do this. The most common way is to use gloves or gauntlets. It is generally recognised that gloves and gauntlets are the most vulnerable parts of an isolator. They are the vital barrier between the hands of the operator and the critical process. It is essential that the operator feels as comfortable as possible with the gloves and that the best attainable manual dexterity is achieved. Where toxic or harmful substances are being handled it is important that materials used in the glove or gauntlet construction provide an effective barrier. The different types of material are listed to help you choose the best or most appropriate one for your particular application. Methods and mechanisms for changing gloves or gauntlets without breaking containment are also described.

It is possible to manipulate materials inside the isolator remotely, without direct operator intervention; thus reference is made to robots which are classed as another form of access device, and hence the title for this chapter.

Once you have narrowed down the choices of isolator configuration, transfer device and access device for the particular project, the next step may be to define the room or location where the isolator is to be sited.

The chapter on **Siting and clothing** should provide enough information on how to construct the room with the appropriate surface finishes, change areas and ventilation requirements. It should be noted that an isolator needs to be in a dedicated area. It is not acceptable to place an isolator just anywhere, in an uncontrolled environment, as it will present too great a risk to the product in the event of a failure. This principle of siting in a dedicated zone will convey an impression to any untrained staff that there is something 'special' about that unit. It will also emphasise to trained staff operating in the unit that the work they do is 'special' or different and needs different and dedicated conditions. Guidance on the type of clothing required by operators is also provided here, as correct clothing will help maintain the correct environment and give the necessary product and operator protection.

The next chapter requiring attention is about **Cleaning, decontamination and disinfection**. It will be necessary to have a well-validated set of procedures in place that ensures you start off with an adequately clean isolator and that it is regularly cleaned to stay that way. The concepts of sterility and sterilisation can cause confusion when isolators are used for preparing patient doses for immediate use rather than for larger scale batch processing. This chapter helps put in perspective the term 'disinfection' and how this differs from 'sterilisation'. Guidance is provided on how to develop a suitable decontamination procedure for your system, whether it be liquid disinfection or gaseous sanitisation.

Now that we have a better understanding of the basics of isolators, we need to understand how to measure how well the system is performing. One way of doing this is by **Physical monitoring**. There are many different tests which can be done to ensure the air is being controlled properly, the filters are performing well and the pressure difference between the controlled workspace and the background environment is being maintained. Physical monitoring should take place both continuously, by checking on-line instruments for differential pressure and air volume change rates, and periodically, by a separate programme of tests. We should now be gaining in confidence with our new isolator. So far, there has been no mention of isolators that leak. Can it be true that comments made about isolators leaking are correct? Surely an isolator contains a controlled workspace, so how can it be controlled if it is full of holes?

Well naturally, manufacturers would not make an isolator that is so bad. They do however admit that it is not possible to make one that is absolutely leak tight! This doesn't mean we have a system out of control. Compare an isolator to an open fronted laminar flow cabinet. Which would have the greater leak rate?

Leak testing is such an important and well-publicised part of physical monitoring that we have decided it warrants a chapter of its own. The extent to which your isolator leaks will determine whether your isolator is likely to be successful in controlling contamination of and from your critical operation. However simple or complex your system, you will need to know just how much your isolator is leaking. Leak detection and leak measurement tests allow you to establish the 'normal' leak rate for your isolator and then, most importantly, to observe if this rate changes over time. The chapter will give you guidance on the different tests that you can use and suggests a number of ways in which 'leak rate' can be defined. If the concept of a 'single hole equivalent' (SHE) is new to you, then there is something in this 'calculation orientated' chapter for you!

Physical tests, including leak tests, can only serve to predict that your isolator will provide the necessary microbiological performance. This can only be truly verified by **Microbiological testing.** The detection of any microbial contamination in the air, on isolator surfaces, on material surfaces and as a result of operations is vital, and tests should be performed frequently and routinely. It should always be a suitably qualified and experienced person who reviews the microbiological monitoring programme.

Before you bring your isolator into use you will need to satisfy yourself that it will perform to expectation in every aspect. The comprehensive chapter on **Validation** should be of help to you in this and may demystify the meanings of 'DQ, OQ, IQ and PQ'. Remember that validation starts at the planning stage and everybody involved in the procurement, specification, design, installation and commissioning of an isolator must have an understanding that it is not a process that is 'just tacked on at the end'. If you have an operation that is regulated, then it must be fully validated.

We have already mentioned compliance with **Standards and guidelines.** There is a whole chapter devoted to all the standards and guidelines that might be relevant to pharmaceutical isolators. These include clean air standards, filter standards, biotechnology standards and of course the European GMP (good manufacturing practice). The chapter starts with a very brief explanation of British, European and International standards and some of the drafting stages.

If some of the words and phrases in this book are new to you, or need clarifying, then perhaps you will take a moment to look up their definitions in the **Definitions** chapter. The authors have sought to utilise or adapt definitions from existing standards and guidelines wherever possible so as to support those standards and guidelines and avoid confusion. Indeed it would be a great step forward if everybody involved in isolators were to be consistent and always use the same definitions.

A series of *Appendices* provides more detailed information on aspects that are touched upon in the main chapters. The HSE/MCA Table of Best Practice sets out the additional measures that are required for product protection or operator protection respectively depending on whether positive pressure isolators or negative pressure isolators have been chosen for use with cytotoxics. No one can dispute the importance of operator **Training**. How do you start? How do you form a good training programme? Where can your staff go to learn about isolators? Can you check on their understanding? How do you document their training? The training checklist should provide you with a number of ideas. **Stainless steel for isolators** gives detailed information about what is the principal material of construction for isolators. You already know which grade to specify, but do you know why? Most if not all isolators use high efficiency particulate air (HEPA) filters to ensure a supply of air of the necessary quality and to remove contamination from exhaust air. But do you know anything about *HEPA filtration mechanisms* whereby a filter has a higher efficiency for particles both larger and smaller than the most penetrating particle size (MPPS) particle? Do you have any idea of the size of hole that is revealed when you detect a leak with DOP (dispersed oil particulate)? The *SHE calculation for DOP* shows how to derive an approximation of the actual size of such a leak. HEPA filters of course can only remove particles. If you need to remove vapours from the exhaust of your isolator, you may wish to use **Carbon filters**, in which case you will want to know more about how they are manufactured, how they work and what you can expect from them

You should by now be in little doubt that buying and setting up an isolator needs special consideration. It is still a straightforward process. We are not trying to dishearten you here, but as for most things of value, it is necessary to follow a careful planning process. Without the close proximity and assistance of a local quality control expert with experience of isolators, you may need to enlist the help of a specialist consultant or manufacturer.

Have we now done all we need to, to have a well-designed, well-controlled and well-maintained system, knowing about leaks and cleaning, with regular revalidation in place? No matter how comprehensively we have adopted the recommendations of this guide, there is still a most serious risk that things are going to go wrong. Without a team of adequately trained staff, working in a good environment, feeling they are part of a well-respected team of people, it is virtually certain that disaster will be just over the horizon. So perhaps the most important of the Appendices is the one on **Training**. After all this, you should then be in a strong position to survey all you have done and say ‘Yes, my system is in control'. If you are a manager your responsibility is highly significant. If you are any other member of the team then you are still extremely important as it depends on you to ensure every precaution is taken to make sure that patients are not compromised as a result of your actions.

The development of this guide has been through the collaborative efforts of a number of experts and specialists. These deliberations have produced the specialist chapters and appendices in the book. It is the wish of the authors that you benefit from the guidance they offer. Suitable solutions for your facility should be found but it is important to remember that technology and regulatory aspects are changing all the time.

The Pharmaceutical Isolator Working Party hope you will find this book of value. Further information and advice will be obtainable from a number of sources, such as the Pharmaceutical Isolator User Group and the ongoing series of specialist conferences.

Abbreviations

ACDP	Advisory Committee for Dangerous Pathogens (UK)
AHU	air handling unit
API	active pharmaceutical ingredient
AS	Australian Standard
BCG	Bacillus Calmette–Guérin
BI	biological indicator
BS	British Standard
BSI	British Standards Institute
CD	Committee Draft (ISO)
CEN	European Committee for Standardization
cfu	colony forming unit
cGMP	current good manufacturing practice
CIBSE	Chartered Institute of Building Services
CIP	clean in place
CIVAS	Central Intravenous Additive Service
COSSH	Control of Substances Hazardous to Health
DIS	Draft International Standard (ISO)
DOP	dispersed oil particulate
DPTE	double-porte de transfert étanche
DQ	design qualification
ECGMP	European Community Good Manufacturing Practice
EN	European Standard (Norm)
EPDM	ethylene propylene diene monomer
ESD	electrostatic discharge
FAT	factory acceptance tests
FDA	Food and Drugs Administration (USA)
FDIS	Final Draft International Standard (ISO)
FDS	functional design specification
FMEA	failure mode and effects analysis
FQT	factory qualification testing
GA	general arrangement (drawing)
GAMP	good automated manufacturing practice
GDP	good distribution practice

GLP	good laboratory practice
GMO	genetically modified organism
GMP	good manufacturing practice
GMPI	Good Manufacturing Practice Institute
HACCP	hazard assessment of critical control points
HAZOP	hazard and operability studies
HCT	high containment transfer
HEPA	high efficiency particulate air
HSE	Health and Safety Executive (UK)
HVAC	heating ventilation and air conditioning
IEC	International Electrotechnical Commission
IMS	industrial methylated spirits
IP	ingress protection
IPA	isopropyl alcohol
IQ	installation qualification
ISO	International Standards Organization
ISPE	International Society of Pharmaceutical Engineers
LCT	life cycle testing
MHRA	Medical and Healthcare products Regulatory Agency (UK)
MPPS	most penetrating particle size
OEL	occupational exposure limit
OOS	out of specification
OQ	operation qualification
P&ID	pipework and instrumentation diagram
PC	personal computer
PD	Published Document (Standards)
PDA	Parenteral Drug Association (USA)
PIC/S	Pharmaceutical Inspection Cooperation Scheme
PLC	programmable logic controller
PN	parenteral nutrition
PQ	performance qualification
Ra	Roughness average
RH	relative humidity
RTP	rapid transfer ports
RTV	room temperature vulcanising
SAB	Sabouraud dextrose agar
SAL	sterility assurance level
SAT	site acceptance testing
SC	Sub-Committee of the TC (ISO)
SHE	single hole equivalent
SIP	sterilise in place

SOP	standard operating procedure
STEL	short term exposure limit
TC	Technical Committee (ISO & CEN)
TPN	total parenteral nutrition
TQT	type qualification testing
TR	Technical Report (ISO)
TSA	tryptone soya agar
UCL	upper confidence limit
ULD	upper limb disorders
ULPA	ultra low particulate air
URS	user requirement specification
VMP	validation master plan
VP	validation plan
WD	Working Draft
WFI	water for irrigation or injection

1

Isolator applications

Can a pharmaceutical isolator be useful in your place of work? This first technical section of the book summarises the many different ways in which an isolator can be used. The roles are varied because the configurations of an isolator are equally varied. It is likely that new ways of using isolators will emerge as time passes. Current applications are identified here.

Introduction

The concept of isolators originated with glove boxes developed for containment in the nuclear processing industry. Not long afterwards, flexible enclosures, which became known as isolators, evolved for specific pathogen-free and germ-free animals. From these two starting points, applications grew to the point where the use of isolators is now widespread in medicine, research and many industries. They can be built in a wide range of materials, sizes and configurations. For any process that might benefit from a controlled environment, i.e. where clean air or containment is required, there will certainly be an isolator solution that is already available or that can be readily developed.

1.1 Applications

The following summary of applications starts with applications outside the pharmaceutical area and continues with the pharmaceutical applications that are the subject of this book. In some of the applications, a high level of containment is required; in others, none. In some, the quality of the internal environment is the vital criterion; in others, it is of no importance. Sometimes a combination is required. Each application therefore needs to be assessed to determine the level of containment and the quality of the internal environment, and then engineered accordingly. Manufacturers of isolators should be able to justify their design in relation to the application.

1.1.1 Non-pharmaceutical applications

1.1.1.1 The food industry

Isolators are used in the production of food and drink with the following objectives:

- to protect raw and finished goods from microbiological contamination during processing;
- to reduce spoilage by processing in a controlled atmosphere or inert environment;
- to increase flavour and shelf-life by excluding moulds;
- to protect operators from allergens.

1.1.1.2 Hospitals

Patient isolators are used for protecting severely immunocompromised patients from microbiological contamination. They can also be used over patients with highly infectious diseases to protect staff from infection.

1.1.1.3 Medical device manufacture

Isolators may be used for the assembly of products such as surgical implants prior to sterilisation or for manipulation after sterilisation. This application is the subject of prEN 13824: Sterilisation of medical devices.

1.1.1.4 Microbiological safety cabinets

Class III microbiological safety cabinets and glove boxes are used for handling highly dangerous pathogens. These are never described as isolators. Their design and testing is covered by BS EN 12469: Biotechnology – Performance criteria for microbiological safety cabinets.

1.1.1.5 Animal laboratories

Isolators are used for containing and handling laboratory animals under study to protect them from infection and infestation, and to provide them with clean and comfortable surroundings. Containment may be required to protect technicians from exposure to animal allergens and pathogens.

1.1.1.6 Nuclear engineering

Glove boxes, which are in effect isolators, are required to provide a very high level of containment with, in addition, a suitable amount of lead shielding against radiation.

1.1.2 Pharmaceutical industry applications

1.1.2.1 Primary manufacture

Applications include:

- keg sampling;
- weighing and dispensing of bulk active pharmaceutical ingredients (APIs);
- mixing and blending of bulk APIs;
- bulk charging and discharging of reactors, blenders, granulators and similar equipment;
- drying and filtering;
- milling;
- explosion-free handling.

1.1.2.2 Secondary manufacture

Applications include:

- keg sampling;
- cleaning of components;
- weighing and dispensing;
- crystallisation;
- micronising;
- blending;
- granulation;
- drying;
- tablet compression;
- tablet coating;
- sterile liquid operations
  - — diluting and mixing
  - — stirring and homogenising;
- aseptic filling (packaging) machines for liquids into ampoules, vials and syringes;
- containment of the filling (packaging) machines of hazardous

liquids and powders into vials, capsules, blister packs and inhalation devices;
- blow–fill–seal operations;
- aseptic access to depyrogenation ovens, autoclaves, freeze dryers (lyophilisers) and vacuum dryers;
- any other aseptic handling of terminally sterilised components.

Volumes of materials and components can be large, and transfer methods need proper consideration. Isolators may be used to supplement cleanroom production areas.

1.1.2.3 Fine powders

Applications include operations where protection of both operator and product is required. Low air movement can be an advantage for weighing. Granite slabs and shrouds can be incorporated into the isolator for high accuracy balances.

1.1.2.4 Contained and aseptic transfers

There is frequently a requirement to transfer materials and components from one controlled area to another:

- transfer between processes, e.g. transfer of vials from a depyrogenation oven to a filling machine isolator;
- aseptic transfer from reactors;
- transfer of sterile components for aseptic assembly, e.g. preparation of TPN (total parenteral nutrition);
- contained and/or aseptic reactor charging and discharging.

1.1.2.5 Sterility testing

For pharmaceutical manufacturers, this is perhaps the most exacting of all sterile applications as a false positive can result in the rejection of a very high value production batch. Speed of processing and a high capacity are important factors as are sterile transfers from the production process to the sterility test isolator. A half-suit system may facilitate sustained activity by the operator.

1.1.2.6 Research

Isolators are used for pilot scale production and on a smaller scale for weighing, analysis and many of the other activities that are carried out under the general heading of research. Isolators can provide containment for compounds of unknown toxicity and potency.

1.1.3 Hospital pharmacy applications

1.1.3.1 Licensed hospital sterile production units and aseptic preparation units

Isolators are used for:

- Central Intravenous Additive Service (CIVAS) with antibiotics;
- Central Intravenous Additive Service (CIVAS) without antibiotics;
- preparation and dispensing of cytotoxic medicines;
- compounding of parenteral nutrition (PN or TPN) or enteral nutrition solutions;
- sterility testing of sterile medicines;
- manipulation of monoclonal antibodies and gene therapy products;
- any other aseptic drug reconstitution, and syringe, vial, bag or device filling.

Batch sizes handled are relatively small compared to industry, response times for individual patient prescriptions more critical, and shelf-lives anything from 'for immediate use' to commercial batch storage. Hospital applications tend to be served by a variety of standard isolator designs whereas isolators for industry are likely to be customised for specialised applications.

1.1.3.2 Radiopharmacy

Radiopharmaceutical isolators are used for:

- preparation of radiopharmaceuticals for nuclear medicine;
- blood labelling.

1.2 Summary

Traditionally, many of the above applications have been carried out using cleanroom technology. The use of isolators has enabled the operator, who is the largest source of contamination, to be separated from the process. Sporicidal gas generators have enabled high levels of bioburden reduction to be achieved and validated. However, the full cycle including aeration can take some time and liquid sanitisation remains a viable option where a rapid response time is important.

Numerous cleanroom environments are now being re-engineered, with large bottling plants being contained in fully validated isolators.

Hazardous chemicals and microbial cultures are more commonly encountered and operators need assurance of their personal safety. Isolators can be designed to provide high containment as well as high microbial decontamination when sporicidal gassing is used.

Table 1.1 shows the different types of isolator that can be used for some typical applications. GMP and hazard control should be in accordance with local regulatory requirements.

Table 1.1 Typical isolator applications

Application	*Positive pressure*	*Negative pressure*	*Flexible film*	*Rigid isolator*[e]	*Laminar flow*[b]	*Turbulent flow*[a,d]	*Gas sanitised*
Aseptic Preparation							
Parenteral nutrition	✓		✓	✓	✓	✓	
CIVAS	✓		✓	✓	✓	✓	
Cytotoxic	✓	✓	✓	✓	✓	✓	
Radiopharmaceutical		✓		✓	✓	✓	
Aseptic Batch Production							
Parenteral nutrition	✓		✓	✓	✓	✓	✓
CIVAS	✓		✓	✓	✓	✓	✓
Cytotoxic	✓	✓	✓	✓	✓	✓	✓
Gene therapy	✓	✓	✓	✓	✓	✓	✓
Live virus		✓	✓	✓	✓	✓	✓
Blood products	✓	✓	✓	✓	✓	✓	✓
Industrial							
Sterility testing	✓		✓	✓	✓	✓	✓
Powder weighing		✓	✓	✓	✓	✓	
Micronising		✓	✓	✓	✓	✓	
Keg sampling		✓	✓	✓		✓	
Reactor loading		✓		✓		✓	
Reactor unloading		✓		✓		✓	
Filling Lines[c]							
Pharmaceutical	✓	✓	✓	✓	✓	✓	✓
Medical products	✓	✓	✓	✓	✓	✓	✓

[a] Some turbulent flow applications may require an inert atmosphere.
[b] Laminar flow can be total or local.
[c] Filling line isolators are usually bespoke and therefore the subject of a design specification.
[d] Turbulent flow isolators should avoid stagnant air pockets.
[e] Rigid isolators can be painted mild steel, stainless steel or plastic depending on application.
Ticks show permissible applications. Configurations which are not ticked and so not recommended in this table may still be appropriate in certain applications.

2

Design

Unless the isolator is correctly designed it is unlikely that it will enable the most efficient or safe working practices, or optimal performance in the working environment. There are a large number of considerations that need to be made when planning to use an isolator. The appropriate selection of the factors in this section will probably be the most important in the eventual successful implementation of isolators in operational practice.

Scope

This chapter characterises pharmaceutical isolators, outlines the most common configurations, describes the basic design parameters to which they may be expected to conform and sets out the considerations that apply. It is not intended to be restrictive or exhaustive but to act as a guide for the factors that need to be considered when a pharmaceutical isolator is specified as part of a design brief.

2.1 Definitions and terminology

Where possible, the definitions and terminology used in this chapter are the same as those used in Draft ISO 14644-Part 7: Separative devices (clean air hoods, glove boxes, isolators and mini-environments). However, Draft ISO 14644-7 is not application-specific; therefore additional application-specific definitions and terminology are taken from other standards and guidelines such as the EC GMP.

2.2 Characterisation of a pharmaceutical isolator

A pharmaceutical isolator is a separative device that separates a pharmaceutical process or activity from the operator and the surrounding environment. This can be for the purpose of:

a) providing a classified clean or classified aseptic environment for a process or activity and protecting it from microbial and/or non-viable contamination arising from the operator and the surrounding environment. This is referred to as product protection;
b) protecting the operator and the surrounding environment from hazards arising from the process or activity. This is referred to as operator protection or containment;
c) protecting the product from contamination generated by other products or processes, either at the same time or during earlier operations. This is referred to as protection against process-generated contamination or cross-contamination;
d) any combination of a), b) and c).

A pharmaceutical isolator can also be used for non-pharmaceutical applications with similar requirements, such as medical devices.

A pharmaceutical isolator system comprises four main elements:

(i) The controlled workspace
This is the defined volume that is created by using a combination of aerodynamic and physical means of separation, in order to achieve the necessary assurance of maintaining separation, as described in the 'separation continuum' in Draft ISO 14644-7 Annex A.
(ii) The transfer device(s)
This is the means whereby materials are transferred in and out of the work zone. There is a range of transfer devices including simple doors, air-purged transfer chambers, and double door transfer ports. These are defined and classified in Draft ISO 14644-7 and described in Chapter 3: Transfer devices, which also gives guidance on the suitability or otherwise of transfer devices for particular applications.
(iii) The access device(s)
This is the means whereby the activity or process in the work zone is carried out. Access devices include gloves and gauntlets for the operator and remote controlled robotic devices. Typical access devices are described in Draft ISO 14644-7 and in Chapter 4: Access devices.
(iv) The decontamination system. This is the means of decontaminating the isolator itself and materials entering and leaving it.

The principal methods of decontamination are:

- Surface decontamination by liquid cleaning combined with sanitisation or disinfection by spraying, swabbing or dunking. This is a relatively quick process that takes 1–10 min.

- Sporicidal gassing with bactericidal, fungicidal and sporicidal agents including 'fumigation' with formaldehyde. This gives a greater level of sterility assurance than liquid cleaning but can take 1–10 h.
- Clean In Place (CIP) by mechanical or automated cleaning systems.
- Sterilise in Place (SIP) by heat or other sterilising methods. This would normally only apply to interfaces with the isolator or to process equipment inside it.

2.3 Design considerations

The detailed engineering design of an isolator should reflect the intended application of the isolator and the chosen decontamination method. Applications will require considerations of the following aspects:

2.3.1 Product protection

An isolator is provided for the manipulation of non-hazardous sterile materials. The products present a low risk to the operator, but must be handled under aseptic conditions to maintain sterility and minimise any contamination. A degree of operator protection minimises risk to the operator.

Example 1 (hospital) TPN compounding, preparation of intravenous additives or reconstitution of sterile dry powder drugs before administration. A positive pressure isolator would probably be specified. The method of decontamination might be sanitisation by spraying or swabbing.

Example 2 (industrial) Filling of sterile products for injection into vials, bags, etc., under controlled aseptic conditions. A positive pressure isolator would probably be specified. The decontamination system might be sporicidal gassing. Different transfer devices might be specified to suit different aspects of the application, e.g. material entry, material exit and waste removal.

2.3.2 Operator protection (containment)

An isolator is provided for the manipulation of non-sterile hazardous materials. The products present a risk to the operator, but do not require aseptic handling.

Example 3 (industrial) Handling cytotoxic products that are subsequently terminally sterilised. A negative pressure isolator would probably be specified.

2.3.3 Operator protection (containment) and product protection

An isolator is provided for the manipulation of hazardous sterile materials. The products present a risk to the operator and also must be handled under aseptic conditions to maintain sterility and minimise any contamination.

Example 4 Formulating, preparing, compounding and filling of cytotoxic materials. Either a positive or a negative pressure isolator may be specified. If a positive pressure isolator is specified, then additional measures may be required to achieve the necessary level of operator protection. If a negative isolator is specified, then additional measures may be required to achieve the necessary level of product protection. A risk assessment should be carried out to assist with the decision.

Double-walled isolators may also be specified for this application. They have a positive pressure critical zone surrounded by a negative pressure zone in the double wall. Such isolators are likely to be expensive and in any event the greatest likelihood of loss of containment is through single skinned components such as gloves or sleeves. It is therefore necessary to engineer out all single-skinned components.

2.3.4 Protection against process-generated contamination

An isolator is provided to minimize or eliminate process-generated contamination. This may be either from the process to other parts of the work zone or from one process to the next.

Example 5 Unpacking from fibre-shedding outer wrappers (e.g. paper) where particulate contamination of the product or medical device is undesirable. Unidirectional or laminar flow in a positive or negative pressure isolator would probably be specified.

Example 6 If an isolator is to be used for a sequence of products, then it should be subjected to a validated cleaning process. Regulatory authorities advise strongly against the use of one isolator where cross-contamination represents a risk and would normally expect to see separate dedicated facilities.

2.3.5 Radiopharmacy

An isolator is provided for the preparation of radiopharmaceuticals and similar activities in nucleur medicine. This is similar to section 2.3.3 in that aseptic conditions are required but there are additional ionising radiation hazards. Protection from radiation hazards requires the containment of radiation-emitting particles or gases and also radiation shielding to protect the operator from direct radiation emissions. The advice of the local radiation protection adviser must be sought in determining the level of radiation protection and shielding required.

Example 7 Isolators for technetium radiopharmaceuticals. A negative pressure isolator ducted to atmosphere would probably be specified. In addition, radiation shielding to a defined lead equivalent thickness should be specified for any area of the isolator where this might be considered necessary.

Example 8 Isolators for blood labelling. A negative pressure isolator ducted to atmosphere situated in a room that is totally separate from any other aseptic processing would probably be specified. This is to eliminate the possibility of cross-contamination from blood-borne pathogens in the blood being handled. Protection from radiation hazards would be as in Example 7. In addition, radiation shielding to a defined lead equivalent thickness should be specified for any area of the isolator where this might be considered necessary.

2.4 Construction

2.4.1 General

Isolators should be constructed from materials and components that are durable, capable of maintaining a good air-tight seal, non-corroding and resistant to agents selected for decontamination. Construction materials can include flexible film, stainless steel, coated steel, glass and a wide range of plastics. All internal surfaces should be accessible to the operator for decontamination purposes, preferably without needing to open the isolator. Supporting structures and all external surfaces should also be easy to clean, and corrosion-resistant.

2.4.2 Main components

The main components of an isolator are the enclosure for the controlled workspace, the transfer devices, the access devices such as gauntlets or gloves and sleeves, the air handling and filtration system, the controls and monitoring devices, and any ports for testing or services.

2.4.3 Flexible film

This may be used as a relatively inexpensive material for forming a large enclosure or envelope of an isolator designed for applications including aseptic dispensing, sterility testing, research and pilot-scale production. A rigid external framework will normally support the flexible film envelope. Floors or work surfaces made of flexible film are not recommended unless the level of usage is to be extremely light.

2.4.4 Stainless steel

This may be used to form the enclosure of isolators for applications where there is a high rate of wear and tear. It may also be used for transfer chambers and similar parts. All parts need to be constructed so that cleaning can be carried out thoroughly and easily. This can be assisted by incorporating radiused (rounded) corners and smooth edges. Welded joints should be free from crevices and non-porous. The standard of internal finish should be specified.

- 304 grade stainless steel is generally specified only for non-product-contact areas such as the enclosure itself; however, 316L grade stainless steel is increasingly being specified for these areas.
- 316L grade stainless steel is generally required by the pharmaceutical regulatory authorities, such as MHRA (Medical and Healthcare products Regulatory Agency)/FDA (Food and Drugs Administration), for product contact areas such as containers, vessels and pipework.

It is strongly recommended that a certificate of conformity of the grade of stainless steel should be obtained.

Further information on stainless steel for isolators is given in Appendix 3: Stainless steel for isolators.

2.4.5 Coated mild steel

This may be used in relatively light applications where non-corrosive and non-abrasive materials are to be handled.

2.4.6 Rigid plastics

These may be used in relatively light applications. Users should be aware of possible hazing, crazing or cracking when plastics are exposed to chemicals such as solvents. It should be noted that the use of plastics could give rise to a release of toxic fumes in the event of a fire.

2.4.7 Windows in rigid isolators

Opening windows may be fitted for access to equipment. They should form an airtight seal when closed. They must be safety interlocked to any moving machinery inside the isolator. Top-hinged windows should be supported safely when open. Windows may be made from safety plastics or safety glass.

2.4.8 Pneumatic and other gas services

Pneumatic door seals or gas services which might leak or discharge into the internal work zone of the isolator should have suitable in-line filtration for the air or gas supplies. It should be noted that such filtration might not remove contamination introduced at the assembly or re-assembly stage.

If an isolator is designed to be sealed, for example as part of a sporicidal gassing cycle, a leak from a pneumatic service may cause overpressure which could result in an explosion. Consideration should be given to the provision of a suitable safety device.

2.5 Filtration

2.5.1 General

HEPA filters remove aerosols and viable and non-viable particles from air passing through them. For air supplied to isolators, they are used to provide the required level of air cleanliness in controlled workspaces. For air leaving isolators, they are used to protect personnel and the external environment from undesirable contamination that has arisen inside. Other methods for the removal of particulate contamination from air supplies may be developed in the future. As with HEPA filters, these would need to meet the required performance of the application.

There are a number of standards that apply to the manufacture and testing of HEPA filters and to the *in situ* testing of HEPA filters once installed. Some national standards are being phased out as the new international standard which covers *in situ* testing, Draft ISO 14644-Part 3, is introduced, but at the time of writing this is still not in its final version. The situation is further complicated by a lack of correlation between the filter manufacturer's standards and the standards for *in situ* testing. This section sets out to make recommendations as how best to specify HEPA filters for pharmaceutical isolators, given the confusion of standards.

For aseptic processing applications and procedures, the filters must be capable of providing air quality in the critical zone inside the isolator that conforms to EC GMP Grade A.

2.5.2 Filter manufacturers' standards

There are two current filter manufacturers' standards in the UK. BS EN 1822: High efficiency air filters (HEPA and ULPA), is a recent European Standard for HEPA and ULPA (ultra low particulate air) filters. BS 3928 is the British Standard that specifies the sodium flame test method for HEPA filters. BS 3928 remains current as a standard, and some filter manufacturers still use the BS 3928 sodium flame test method. BS EN 1822 appears to be the preferred standard for specifying HEPA filters for pharmaceutical applications and utilises a scan test which gives both overall and local penetration values. The classification is shown in Table 2.1, which includes a note that local values may be agreed between supplier and purchaser. The BS EN 1822 test cannot be reproduced on site for installed filters. The BS 3928 sodium flame test is a volumetric test which does not give a local value. Where a local value is important for the application, e.g. for filters delivering unidirectional air, some filter manufacturers carry out the test specified for the *in situ* testing of installed HEPA filters for testing filters in the factory. On-site testing of

Table 2.1 Classification of HEPA and ULPA filters from BS EN 1822-1:1998 Part 1: Classification, performance, testing, marking

Filter class	*Overall value*		*Local value*[a]	
	Efficiency (%)	*Penetration (%)*	*Efficiency (%)*	*Penetration (%)*
H 10	85	15	–	–
H 11	95	5	–	–
H 12	99.5	0.5	–	–
H 13	99.95	0.05	99.75	0.25
H 14	99.995	0.005	99.975	0.025
U 15	99.999 5	0.000 5	99.997 5	0.002 5
U 16	99.999 95	0.000 05	99.999 75	0.000 25
U 17	99.999 995	0.000 005	99.999 9	0.000 1

[a] Local values may be agreed between supplier and purchaser.

installed filters with a BS 3928 sodium flame test is possible but is rarely carried out. It should be noted that volumetric leak tests are much less rigorous than scan leak tests.

BS EN 1822 states that the mean particle diameter of the test aerosol for testing filter efficiency and penetration shall correspond to the MPPS for the filter medium at the nominal volume flow rate. MPPS is explained in Appendix 5: HEPA filtration mechanisms. Manufacturers take the nominal volume flow rate as meaning the rated volume flow rate for the particular filter. For cleanroom filters, filter manufacturers test and report penetration at a face velocity that gives an air velocity of 0.45 m s^{-1} at the working position. This is in line with the EC GMP guidance value of 0.45 m s^{-1} for 'laminar air flow systems'. High capacity filters are rated at higher volume flow rates and are tested accordingly. Where designers specify 'off the shelf' filters, they should ensure that their designs allow filters to operate at their rated flow rate. Where they specify special filters, they should ensure that filters are manufactured and tested to give the necessary performance at the flow rate required by the design.

2.5.3 Standards for the *in situ* leak testing of installed HEPA filters

In situ leak tests of installed HEPA filters are to test for filter and filter seal leaks arising during transportation, installation or operation. They do not confirm the filter efficiency.

BS 5295: Environmental cleanliness in enclosed spaces, covers *in situ* testing of installed HEPA filters, but has largely been withdrawn in anticipation of the phased introduction of the various parts of BS EN ISO 14644: Cleanrooms and associated controlled environments. British Standards have published a guideline PD 6609-2000: Environmental cleanliness in enclosed spaces – Guide to test methods. This is a temporary document which reinstates, as a guideline, the *in situ* leak testing method of BS 5295 pending the finalisation of Draft ISO 14644-Part 3: Metrology and test methods.

Neither the tests of PD 6609, nor those of draft Draft ISO 14644-3 replicate the manufacturer's factory test because:

- the test particles are different as shown in Table 2.2;
- in BS EN 1822 the percentage penetration is of the number of particles in the challenge, whereas in PD 6609 and Draft ISO 14644-3 using an aerosol photometer, it is of the total mass of the particles in the challenge.

The information in Table 2.2 is a summary of detailed information given in comprehensively written standards. These must be referred to before decisions are taken.

2.5.4 Specification of HEPA filters

HEPA filters must meet the requirements of the application. Therefore, when purchasing or specifying HEPA filters, the following factors are as important as the classification in BS EN 1822:

- the maximum intended filter face velocity or volume flow rate (This is the worst case.);
- a requirement for the filter manufacturer to state the MPPS at the intended face velocity or volume flow rate;
- the filter case size: length, width and depth and effective face area;

Table 2.2 Standards for leak testing: a summary of detailed information on test particle size and penetration

Test	*Test particle size*	*Maximum overall penetration*	*Maximum local penetration*
BS EN 1822 Manufacturers'	MPPS (typically 0.1–0.2 μm)	0.005% (for H 14)	0.025% (for H 14)
BS 3928 Manufacturers' (UK only)	0.02 μm to 2 μm Mass median size about 0.6 μm	N/S in BS 0.005%, 0.003% or 0.001% in practice	N/A
PD 6609 Installed filters (UK only)	More than 20% by mass less than 0.5 μm More than 50% by mass less than 0.7 μm More than 75% by mass less than 1.0 μm	N/A	0.001%[a] for Classes C, D, E and F[b] 0.01% for other Classes[b]
Draft ISO 14644-3 Installed filters	Mass median diameter typically between 0.5 and 0.7 μm	0.01% in a duct or AHU[c]	0.01%

[a] 0.001% penetration appears to be unique to the UK.
[b] The classes referred to are from BS 5295 (withdrawn). Class F is equivalent to ISO 5. Other classes are ISO 6 and higher.
[c] AHU = Air Handling Units.

- the filter pack size: length, width and depth (The actual area of filter media would be calculated by the filter manufacturer to meet the specification.);
- the filter case material: stainless steel, extruded aluminium, etc;
- the (clean) pressure drop across the filter when in a new or clean state;
- the intended *in situ* leak test method: aerosol source oil, aerosol generation method (pneumatic or thermo-pneumatic) and the leak test acceptance criteria;
- the filter gasket or seal type;
- the filter gasket or seal location: clean side, dirty side or both;
- any out-of-the-ordinary requirements such as high airflows, the presence of specific gaseous sterilising agents, extremes of temperature or extremes of humidity.

Current practice accepts that, for most pharmaceutical isolators, the nearest manufacturers' specification for a suitable HEPA filter is a 'good' H 14 according to BS EN 1822-1. A 'good' H 14 filter will have a local penetration at MPPS $\leqq$ 0.01% instead of MPPS $\leqq$ 0.025% as normally allowed for an H 14. This should ensure that the filter will pass the *in situ* leak test at a local penetration of 0.01%. For filters that are tested to BS 3928, the *in situ* test method and acceptance criteria for overall penetration must be specified.

Either panel HEPA filters or cartridge HEPA filters may be used. Wood, plywood, hardboard or MDF should not be used in the structure of HEPA filters as these materials tend to shed particles and can harbour microorganisms.

2.5.5 Incorporation of HEPA filters into isolator designs

There should be at least one HEPA filter at the inlet of an isolator to ensure air of the appropriate quality inside and at least one HEPA filter on the exhaust. Air-purged transfer devices as described in Chapter 3 must have at least one HEPA filter to maintain the physical barrier of the isolator at all times.

In aseptic processing isolators, if the inlet HEPA filter on the isolator enclosure boundary is at risk of damage during cleaning or processing, this damage may only be detected during *in situ* HEPA filter testing. Double HEPA filtration at the inlet is therefore recommended by regulatory authorities unless the isolator is in a classified cleanroom with a HEPA filtered air supply and is sited so that it can draw air that has not been recontaminated in the room. A second in-line HEPA filter is also recommended where filter changes are required as part of the cleaning process for a product change.

For applications where hazardous aerosols or powders might be generated inside the isolator and the isolator exhausts to the room, two HEPA filters should be fitted in series to the exhaust for additional safety. The primary HEPA filter should be safe to change. Subject to local regulations, it is usually sufficient to have one HEPA filter on exhaust systems ducted to atmosphere. This should be safe to change in order to protect the process, personnel and the environment from contamination during filter changes. Where a safe-to-change filter system is required, the manufacturer should provide a suitable design including a suitable change protocol.

An example of a safe-to-change filter design would be a proprietary filter box that allows the contaminated filter to be removed straight into a bag, with no exposure to the external environment. It should be noted that such a system may not be compatible with the filter mounting arrangement shown in Figure 2.1 (see p. 26).

An example of a safe-to-change filter protocol would be the application of a fixing spray to the dirty side of the filter to seal in the contamination before the filter is removed. Hairspray is sometimes used for this, but it has been suggested that spray paint is better because it is possible to verify the extent of the coverage.

A risk assessment would need to be performed to determine which method is the most appropriate for a particular application.

It should be noted that HEPA filters are only effective in removing particles (liquid or solid) from isolator exhaust air, and not vapours or gases. Vapours or gases might be from liquids that have vaporised or solids that have sublimed inside the isolator, or from particles of these substances deposited on HEPA filters. Where toxic vapours might be present, the isolator exhaust should be ducted to atmosphere or carbon filters of a suitable type should be used (see section 2.5.6).

Double HEPA filtration has the further advantage of increasing efficiency. As an approximation, the penetration of the two filters in series can be multiplied together to give a figure for overall penetration that is orders of magnitude less than for a single filter. This increase in efficiency can only be validated if the penetration of each filter can be tested independently. Another means of improving filter efficiency is to reduce the volume flow rate as explained in Appendix 4: HEPA filtration mechanisms.

Where appropriate HEPA filters should be protected with pre-filters to prolong their serviceable life.

Provision must be made for the *in situ* leak testing of individual HEPA filters, with their seals and mounting frames. For every filter there should be provision for the injection of the test aerosol including measurement of the upstream challenge aerosol concentration and suitable access for downstream scanning or volumetric measurement of the filter efficiency. Measurement of aerosol concentration is usually performed with an aerosol photometer. Unidirectional flow HEPA filters must be face-scanned for leakage. This is because in unidirectional flow the airflow from a leak remains as a concentrated stream in the general airflow. Other HEPA filters should be scanned if practicable, even if this entails some dismantling.

Where scanning is not possible, volumetric overall leak tests with the test aerosol are acceptable provided that location of the test port is such that the air being tested is fully mixed.

2.5.6 Carbon filters

Carbon filters should be used on the exhaust where vapours need to be removed. They should be fitted downstream of the exhaust HEPA(s) and a further suitable filter fitted if there is any possibility of release of carbon particles that might be contaminated with the substance adsorbed. Since carbon filters vary in type and can be specific for certain vapours, it is important to:

- specify the appropriate type of carbon for each application;
- specify and design a sufficient 'dwell time' for air passing through the filter;
- change filters regularly in accordance with the manufacturer's life expectancy for the filter;
- have the manufacturer recommend a suitable test method which should be used regularly.

Carbon filters should not be exposed to HEPA filter leak test aerosols as this may seriously reduce the effectiveness of the carbon filter. Over a period of time, carbon filters may release adsorbed materials back into the air.

Carbon filters with broad spectrum organic adsorption capabilities should be specified for isolators where there is concern about the presence of cytotoxic vapours. It has been suggested that such carbon filters would be effective in adsorbing cytotoxic molecules even if the filters were saturated with alcohol. This is because the larger cytotoxic molecules would displace the smaller molecules of alcohol. This has not yet been verified experimentally, nor has it been confirmed that cytotoxic substances such as cyclophosphamide and 5-fluorouracil exist in the vapour phase at room temperature.

For further information on carbon filters see Appendix 6.

2.6 Pressure regimes

2.6.1 Positive pressure isolators

Positive pressure isolators run at operating pressures which may range from +25 Pa to +100 Pa. Low operating pressures are more difficult to monitor and alarm and are more prone to go negative momentarily, for example when glove/sleeves are withdrawn rapidly. High operating pressures are less comfortable for operators because of the pressure on sleeves, gauntlets and half-suits.

2.6.2 Negative pressure isolators

Negative pressure isolators run at operating pressures which may typically range from a nominal –50 Pa to –250 Pa in the controlled workspace. Low operating pressures (more negative) cause sleeves and half-suits to become less flexible as a result of the higher differential pressure between these flexible items and the controlled workspace. High operating pressures (less negative) can be more difficult to monitor and set alarm levels. They may also be more prone to go positive, for example when a glove/sleeve assembly is inserted rapidly into the controlled workspace.

2.6.3 Containment

Negative pressure isolators must be designed so that containment is maintained in the event of a breach. A minimum figure of 0.7 m s^{-1} breach velocity (inward velocity of air through a breach) with a single glove removed, is widely used and is currently specified in BS EN 12469. As the value of 0.7 m s^{-1} is arbitrary for one particular breach condition, any additional requirements for containment or validation should be agreed between customer and supplier. Lower breach velocities have been successfully tested for containment. If such lower breach velocities are to be specified, then the manufacturer should provide containment validation data. Higher breach velocities are thought to cause containment failure as a result of edge effects (see section 2.7.10).

2.6.4 Pressure differentials

In isolators with transfer chambers or in complex isolator systems where two or more enclosures are linked, the pressure differentials between adjacent compartments should support the predominant requirement. This will either be product protection or product containment. When doors are opened between adjacent compartments, air transfer should be controlled by pressure differentials or by aerodynamic means. Pressure differentials and airflow should also be designed to minimise the effect of leaks.

2.7 Leakage and leaktightness

2.7.1 Summary

All isolators leak to some extent. It is therefore important to understand the possible effects of leaks in all types of isolator, to be able to quantify and rationalise a leakage rate that is acceptable in each case. If the leakage rate is set at too relaxed a level, then excessive contamination might pass through the leak to the detriment of the performance of the isolator. Conversely, if it is set at too tight a level, the cost of the isolator and its ongoing monitoring might be unnecessarily high. The maximum acceptable leakage rate should be determined for every design of isolator in relation to its intended application.

The glove/sleeve and/or half-suit systems are the most likely components of an isolator to suffer wear and tear and thus develop leaks. As they are likely to be in the critical zone, they warrant special attention at the design stage and when monitoring protocols are developed.

For more information see Chapter 4: Access devices, and Chapter 8: Leak testing. Other leakage mechanisms are also described and discussed.

2.7.2 Leakage in positive pressure isolators

In positive pressure isolators, leaktightness is primarily important where sporicidal gassing is used or where hazardous materials are handled. Any leakage from the isolator may be hazardous to the operator and the occupational exposure limit (OEL) for the material must not be exceeded. Sporicidal gassing systems may include provision for isolator leak testing and the maximum acceptable leak rate may be determined during development of the gassing cycle, when the air change rate in the room should be taken into account.

In addition, it is possible for positive pressure isolators to go negative momentarily during the removal of hands from gloves. Therefore, a value for leakage of 1.0% volume per hour (equivalent to a pressure decay of 25 Pa from 250 Pa in 1.5 min) is widely used.

2.7.3 Leakage in negative pressure isolators

In negative pressure isolators, leaktightness is primarily important because contaminated air from the background environment may enter the controlled workspace and compromise the aseptic environment. For turbulent flow isolators with a low air change rate of the order of 180 air changes per hour, an arbitrary value for leakage of 0.25% per hour may be used. For laminar flow type isolators where the air change rate is much greater at around 1800 air changes an hour, a more relaxed value for leakage of 1.0% volume per hour is widely used. The rationale for this is based on the greater dilution and the more rapid removal of any airborne contamination that might enter the controlled workspace of the isolator. This dilution is not maintained if the airflow stops as in an electrical supply failure.

It should be noted that even at moderate operating pressures (depressions) a leak in the isolator envelope could result in a gradually dispersing jet of air moving at several metres per second. This jet will only break up fully when it strikes something solid.

Isolator manufacturers should estimate the rate of entry of contamination at a given leak rate and its rate of removal in the isolator airflow.

2.7.4 High pressure integrity/low hourly leak rate isolators

A further leakage class of 0.05% volume per hour is specified for high containment enclosures such as Class III microbiological safety cabinets and nuclear glove boxes. This leakage class is not normally specified for pharmaceutical isolators.

2.7.5 Gloves, sleeves and half-suits

Leaks in gauntlets, gloves, sleeves and half-suits are likely to present a greater risk than those in the main body of an isolator. This is because these components are:

- the least robust components of the isolator;

- the point of closest contact between the process and the operator;
- even when not in use, they may be close to critical parts of the process.

The donning of gloves and gauntlets may produce a pumping action that can raise the pressure inside them and force trapped air into the isolator.

For these reasons, it is important to:

- specify gauntlets, gloves, sleeves and half suits that are of good quality and have an appropriate resistance to wear and tear;
- specify suitable test protocols for frequent monitoring for leaks.

For more information see Chapter 4: Access devices and Chapter 8: Leak testing

2.7.6 Internal leakage

Internal leakage can occur between different sections of an isolator, for example across internal doors. Designers should consider how such leakage should be tested at validation and revalidation and make suitable provision.

2.7.7 Filter seals

HEPA filter seals may leak. It is possible to mount them in a fail-safe manner as described in BS EN 12469, which states that 'HEPA filters shall be mounted in such a way that no air can bypass the filter medium'. This would ensure that any leakage across a seal does not release background air into positive pressure isolators or release potentially contaminated air from negative pressure isolators. Both situations are shown in Figure 2.1 with 'recommended' and 'not recommended' configurations.

2.7.8 Clean air

To ensure the supply of clean air to negative pressure isolators, designers can adopt the other suggestion in BS EN 12469 that any 'airways containing contaminated air chambers, which are under positive pressure, should be surrounded by internal airways at negative pressure'.

2.7.9 Induction leakage

According to Bernoulli's law, the static pressure of a moving air stream is less than that of the surrounding still air. It is therefore possible that

(a) Positive pressure isolators – supply air

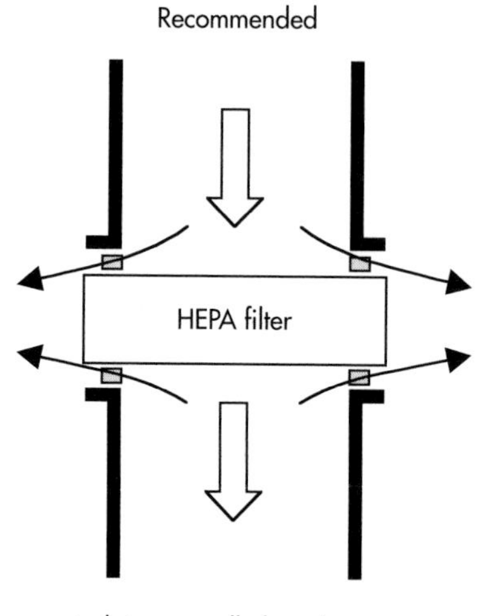

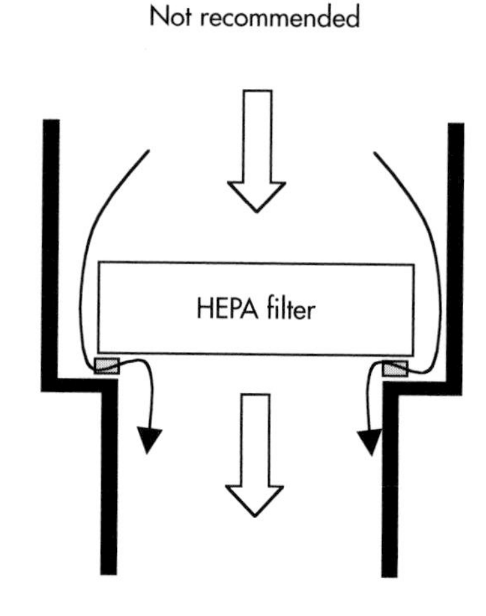

(b) Negative pressure isolators – exhaust air

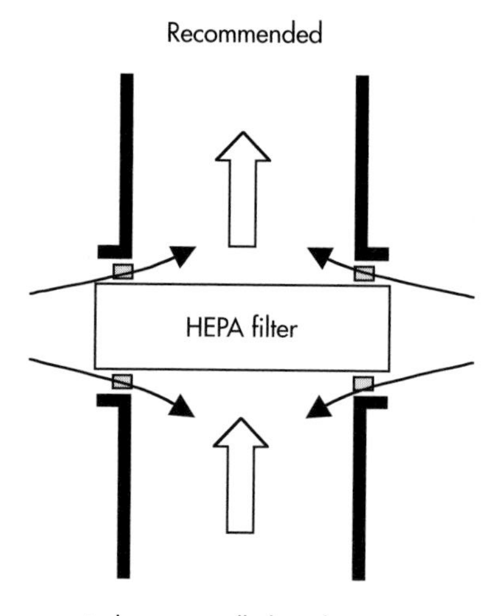

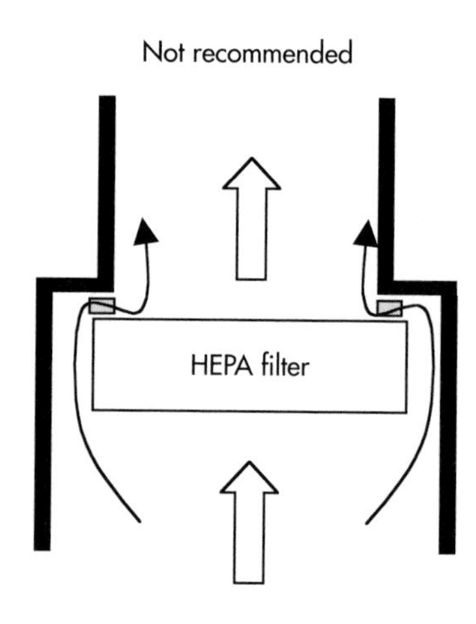

Figure 2.1 HEPA filter mountings.

local air velocities in certain designs of isolator are high enough to create a zone of low static pressure. In a positive pressure isolator, if the normal overall operating pressure is reduced at any time, perhaps as a result of a rapid arm withdrawal, then the combined effect may be a brief local episode of negative pressure. This could produce a momentary inflow of air should there be a leak in that zone. Air velocities that might give rise to this effect should be avoided in isolator design.

2.7.10 Edge effects

Edge effects occur where turbulence and eddies around the edge of an opening produce a localised airflow which is against the general direction of the airflow through that opening. This localised airflow could cause airborne contamination to travel in the opposite direction to that expected. Edge effects can occur at legitimate openings such as those in transfer devices, at accidental breaches such as loss of gloves, and at leakage points. The possibility of such edge effects occurring should therefore be taken into account in the design of openings and the specification of a suitable breach velocity (see section 2.6.3). The absence of edge effects should be verified by means of flow visualisation tests, using a mock-up if necessary.

2.7.11 Macro-leakage

It has been suggested that whilst airflow may prevent the transfer of microorganisms through apertures, macro-organisms such as mites or flies would still be able to enter. To ensure this does not happen during unattended down time, transfer devices such as mouse holes (type A2), should be capable of being sealed. Good housekeeping of the background environment for the isolator should minimise the possibility of macro-leakage.

2.7.12 Measurement and detection of leaks

The process of leak detection, leak-rate measurement and the classification of isolators by leak rate are discussed in detail in Chapter 8: Leak testing.

2.8 Airflow (flow) regimes

2.8.1 Turbulent airflow (non-unidirectional)

Isolator airflow may be turbulent. The air change rate might be as low as 3 air changes per hour in the case of isolators with CIP, such as those that are used for handling bulk quantities of non-sterile potent powders in primary production. The air change rate is normally not less than 20 air changes per hour in the case of isolators that are subject to sporicidal gassing. The entire volume of the isolator should be purged by the airflow, with no stagnant areas or standing vortices. The higher the air change rate, the quicker the purge time for any process-generated contamination and, in the case of negative pressure isolators, in-leakage. Some turbulent flow isolators have an air change rate of 250 air changes per hour.

2.8.2 Unidirectional airflow

Isolator airflow may be unidirectional or laminar. The EC GMP suggests that laminar air flow systems should provide a homogeneous air speed in the range 0.36 m s^{-1} to 0.54 m s^{-1} (guidance value) at the working position. In isolators the air flow velocity is less critical than in open fronted laminar air flow workstations. The average airflow velocity in isolators is normally within the range 0.25 m s^{-1} to 0.5 m s^{-1}. A typical unidirectional or laminar air flow velocity might be 0.4 m s^{-1}. Some specialised applications use velocities as low as 0.03 m s^{-1}. In a unidirectional or laminar air flow system, the uniformity of the air flow velocity should be within a tolerance of $\pm$20% of the mean. A high unidirectional air flow velocity gives quicker purging even when the unidirectional flow is broken up into turbulent flow by, for example, sleeves. A low unidirectional air flow velocity is better for delicate operations such as high accuracy balance weighing and has the further benefits of improving HEPA filter efficiency (see section 2.5.5), reducing power consumption and reducing noise. A typical example of a unidirectional airflow isolator with a downflow air velocity of 0.4 m s^{-1} and a distance from filter to work surface of 0.8 m would have an air change rate of 1800 air changes per hour. It must be emphasised that inside an isolator the value of the unidirectional airflow velocity is less critical than in a conventional 'laminar flow cabinet' due to the separation already provided. Unidirectional flow may be expected to give an increased level of assurance of separation from contaminants for aseptic operations, provided arrangements are made so that they are carried out in freshly HEPA filtered air. See also Appendix 1: 'Handling Cytotoxic Drugs in NHS Pharmacies' HSE/MCA (January 2003) and refer to the EC GMP Revised Annex 1 (May 2003).

2.8.3 Zoned airflow

Isolators may also have a zone of unidirectional airflow within an overall turbulent airflow system. In this case the aseptic operations would be carried out in the zone of unidirectional airflow.

Where a sporicidal gassing or fumigation system is specified, it will connect into the airflow system of the isolator. The way in which this connects will depend upon the design of both the isolator and the gas generator and will vary between manufacturers. Three common systems are:

- gas circulated solely by the gas generator;
- gas circulated by the gas generator with internal supplementary stirrer fans or gas distribution devices;
- gas circulated by the gas generator and the air handling system of the isolator.

2.8.4 Gassing

Gassing or fumigation must be followed by aeration in order to reduce the residual level of the gassing or fumigation agent to a value at which the isolator may be safely opened without exposing operators to levels of gas above the OEL. This is done by exhausting to atmosphere, by chemical neutralisation, or by use of a catalyst. Other important considerations in the integration of a sporicidal gassing or fumigation system with an isolator airflow system are:

1. Provision must be made for the sporicidal gassing or fumigation of inlet and exhaust HEPA filters. Where double HEPA filtration is used, this applies to any filter forming the boundary of the controlled workspace of the isolator.
2. Adsorption of the sporicidal gassing or fumigation agent onto the HEPA filters and materials in the system can make the aeration period protracted due to outgassing.
3. Dead zones in the plenums of the isolator airflow system (or in the work zone itself) when the isolator is sealed off for gassing or fumigation may become zones in which the decontamination process or the subsequent purging is ineffective.
4. Valves or blanking plates to prevent the loss of sporicidal agents during gassing or fumigation must be readily accessible and clearly identified, as must the gassing or fumigation connections.
5. Certain sporicidal gassing or fumigation systems require a pressure sensing line to be attached to the isolator so that the generator may control the isolator pressure during gassing. A suitable connector should be available on the isolator and clearly identified.

2.9 Fan systems

Isolators may have a single fan on the inlet for positive pressure operation, or a single fan on the exhaust for negative pressure operation. This system allows control of the isolator pressure or of the volume flow rate, but not of both independently.

Isolators may have fans on inlet and exhaust so that both the isolator pressure and the airflow rate may be controlled independently. This arrangement, sometimes referred to as 'push–pull', must have a stable control system.

Isolators with more complex airflow systems such as unidirectional flow isolators with a proportion of air recirculating internally may have a combination of inlet fans, downflow (laminar flow) fans and exhaust fans.

Where toxic materials are handled, or large amounts of alcohol are used for surface decontamination, the exhaust from the isolator should be through a sealed negative pressure exhaust duct to atmosphere. The exhaust fan should be at the outer end of the duct. It should be capable of being controlled, preferably from the isolator room or the isolator control system. For safety reasons a local switch should also be provided.

Where the exhaust from the isolator is ducted to atmosphere, care should be taken to allow for the effect of this loss of air from the room to ensure that the room pressure is always positive. This can be done in three ways:

(i) By estimating the volume of the filtered air supply to the room and ensuring this is greater than the isolator exhaust volume. A pressure release mechanism should be fitted to allow surplus air to escape from the room. When the isolator exhaust is switched off, there will be more surplus air and the room pressure will rise. This is the most common method of adjusting for the isolator exhaust.
(ii) By designing sophisticated pressure and airflow volume controls into the air-handling system so that the room pressure is stabilised against changes in exhaust volume.
(iii) By fitting a thimble exhaust to the isolator. The exhaust fan at the outer end of the exhaust ducting is of a sufficient capacity to give the required number of air changes in the room. The thimble allows air to be taken from the isolator exhaust or from the room. When the isolator is switched on the thimble ensures that the isolator exhaust air is taken preferentially. When the isolator is switched off the thimble allows an equivalent volume of room air to be taken instead of isolator exhaust air. This method is commonly used in negative pressure microbiological containment rooms.

All isolator exhaust systems, whether direct or thimble, should be designed so that they are safe in the event of failure. Consideration should be given to the fitting in exhaust ducts of anti-blowback dampers.

These should be sufficiently responsive to ensure that there is no momentary exhaust blowback during abnormal wind conditions or exhaust fan failure. Thimble exhaust systems should be fitted with interlocks so that in the event of the remote thimble fan failing, the isolator cannot discharge to the room.

2.10 Controls, instrumentation, alarms and performance monitoring

2.10.1 Controls

All controls, indicators and gauges should be clearly identified. It should be remembered that the output from any measuring device, and the subsequent handling of the data, may require validation (see Chapter 10: Validation).

The isolator with its transfer devices should be provided with all the controls necessary to ensure its proper function. These can include fan speed controls, remote door controls, light switches and dampers. The level of sophistication should relate to the application.

Where the control of the isolator is by a PC or PLC, the supplier should demonstrate that the control system has been developed, produced and validated within a recognised quality system, for example GAMP.

2.10.2 Instrumentation

Internal operating pressure should be indicated by a calibrated gauge or an instrument with an analogue or digital output. These outputs may be used for continuous monitoring or for control.

Airflow rate should be indicated by a calibrated gauge or instrument. Where an instrument is used it may have an output such as a 4–20 mA analogue signal or an RS 232 or RS 485 digital communication port. This may be used for continuous monitoring or for control.

Partial blockages across any of the principal filters should be indicated by calibrated gauges or instruments. Alternatively, it is acceptable for this information to be deduced from readings on other gauges or instruments.

2.10.3 Alarms

Parameters critical to the safe and proper function of the isolator and its transfer devices should be indicated and alarmed. These must include isolator pressure, airflow rate and mains power.

Alarms should be visual and audible (>95 dB) and have the facility for remote connection, e.g. central control station. There should be provision for testing the correct function of all alarms.

Critical alarms must be 'latched', i.e. they may only be cancelled by active operator acknowledgement and reset. Australian Standard AS4273-1999 requires alarms to remain functional in the event of mains power failure. A risk assessment should be carried out to determine whether an isolator should remain off or switch itself back on after a mains power failure. (Where there is a remote fan, there may be a safety issue.) If it switches itself back on, the alarm should indicate the failure until cancelled by active operator acknowledgement and reset.

2.10.4 Performance monitoring

Additional instruments may be fitted as required to monitor other parameters such as humidity, oxygen content, etc.

Isolators for aseptic applications should be fitted with connections to allow for particle counting.

Isolators for aseptic and other applications may be fitted with connections to allow for microbial air sampling.

Isolators should have facilities to allow for routine leak testing by pressure decay. Such facilities (e.g. shut-off valves or blanking plates) should be clearly identified.

Monitoring connections and devices must not compromise the internal environment of the isolator.

2.11 Ergonomics, lighting, noise, vibration and electrical safety

2.11.1 Ergonomics

Ergonomics and anthropometrics must be considered in the design of isolators. Isolators must be comfortable for a wide range of personnel to operate and work at, either seated or standing, for extended periods. Their use must not require awkward postures or difficult actions, which may give rise to upper limb disorders (ULDs). They must be safe to operate.

2.11.2 Lighting

Useful guidance on lighting is given in the *Lighting at Work* booklet issued by the UK Health and Safety Executive (HSE), which refers to the

Chartered Institute of Building Services (CIBSE) Codes for Lighting and Interior Lighting and in BS EN 12469: 2000.

CIBSE indicates that 500 lux should be the average lighting level and 200 lux the minimum lighting level for the perception of fine detail. For close work, a minimum of 500 lux is recommended. The values are measured at the work surface. These levels are proposed to avoid visual fatigue and should be adequate for safety purposes. The document generally refers to the assembly of electronic components and textile production, but this should also be appropriate for an isolator controlled workspace.

BS EN 12469 states that for microbiological safety cabinets, which might be considered analogous to some pharmaceutical isolators, lighting should be adequate for safe working within the cabinet. Illumination measured at the work surface should be at least 750 lux.

The lighting levels indicated in these guidelines represent the range of lighting levels from which an appropriate level for a pharmaceutical isolator can be selected according to the application.

Light energy is absorbed at different rates dependent on the surface characteristics, so isolators with stainless steel interiors will require stronger light sources than isolators with white interiors to achieve the same effective lighting levels.

2.11.3 Noise levels

Noise levels should be as low as possible and should in any event not exceed 65 dB at 1 m from the worst single sound source.

2.11.4 Vibration

Vibration in applications where low vibration is a critical requirement should not exceed 0.005 mm RMS amplitude in the centre of the work surface between 20 Hz and 20 000 Hz when the isolator is operating normally.

2.11.5 Electrical safety

Electrical systems, including controls, of isolators should be in accordance with IEC 61010-1: Safety requirements for electrical equipment for measurement, control and laboratory use – General requirements. Electrical wiring should be outside the controlled workspace of the isolator where possible. Where electrical enclosures are accessible, they should be to IP 65 to BS 5490 (dust tight and protected against water jets), to allow for cleaning of the isolator.

2.12 Design for decontamination

Pharmaceutical isolators must be designed so that they can be easily cleaned and decontaminated. Material transfer and decontamination should be addressed as part of the system design process. Aspects to be considered are:

1. Compatibility of materials with agents used for decontamination;
2. Smooth internal design with minimal cracks and crevices, smooth welds and corners, optimal surface finish and minimal inaccessible areas;
3. Ergonomics and access, by means of special tools if necessary. Ideally, all internal surfaces should be easily accessible for cleaning without the use of special cleaning tools. If such tools are required then they should be provided as part of the isolator package.
4. Issues relating to health and safety such as operator inhalation and explosion risk;
5. Aspects specific to any gassing system that is selected. Such aspects might include connections, gas distribution, surface exposure, temperature distribution and biological decontamination of HEPA filters.
6. For aseptic processing using liquid sanitisation the transfer device should prevent entry of airborne contamination into the controlled workspace. The liquid sanitisation should prevent the entry of surface contamination on materials to be processed and their containers.
7. For aseptic processing using sporicidal gassing, isolators should be capable of being fully sealed for the gassing process. Where a rapid response is required, it is advantageous to have separate isolators for sporicidal gassing of incoming materials, storage of treated materials and processing. This ensures availability of materials for immediate processing. All the isolators should be gassed regularly and material transfers between the different isolators should use type E or F transfer devices (see Chapter 3: Transfer devices).
8. The smaller the volume of the isolator or the transfer device to be gassed the shorter the gassing phase. The higher the air change rate in the volume to be gassed the shorter the aeration phase of the gassing cycle.

2.13 Design for validation

Pharmaceutical isolators for facilities subject to regulatory compliance should be capable of formal validation. The following is a brief outline of the main points concerning design for validation. A comprehensive summary of all aspects of validation is given in Chapter 10: Validation.

The isolator should be designed to meet the requirements of the User Requirement Specification (URS) and a Failure Mode and Effects Analysis (FMEA), or equivalent, carried out as part of the design process.

Upon completion of the design, a Design Qualification (DQ) should be undertaken to confirm compliance with the URS.

Upon installation, the isolator will require qualification to ensure it has been delivered and installed to the agreed design. This is called Installation Qualification (IQ). It is therefore helpful if the aspects that are to be qualified at the IQ stage, including the design, are clearly documented prior to delivery. IQ includes qualification of the following:

- the equipment itself;
- the factory test report or the report of the factory acceptance tests (FATs);
- the calibration certificates for instruments and gauges on the isolator.

When installation is complete, the isolator will require qualification to ensure that it functions in accordance with the agreed design. This is called Operation Qualification (OQ) and includes physical testing to confirm the following:

- correct function of all controls, instruments and the isolator itself;
- achievement of all performance parameters;
- in the case of sporicidal gassed isolators, gassing cycle development.

Test points for physical testing should be included in the design. It is recommended that the physical test protocol is written as part of the design stage in order to ensure that all the necessary tests points are provided. These must include as a minimum:

- test ports for HEPA filter integrity tests (see section 2.5.5);
- connections for testing isolator pressure in the main chamber;
- connections for testing pressures in transfer chambers and any other compartments where the pressure may be independent of the isolator main chamber pressure;
- connections to measure pressure drop across critical HEPA filters;

- facilities for testing isolator airflow rates and unidirectional airflow velocity;
- facilities for testing leaktightness;
- test points for gas concentration in the case of isolators that are sporicidally gassed.

All gauges or instruments should be capable of calibration.

All controls should be clearly labelled so that their function and the correct function of the isolator can be systematically checked.

When documentation for IQ and OQ is required, it should be produced as part of the design stage to ensure that all necessary performance values, instrumentation and test points are included in the design. Documentation should include appropriate checklists (see Chapter 10: Validation).

Performance qualification (PQ) should be carried out by the user and is the ultimate test of design.

3

Transfer devices

An isolator will retain its integrity provided the seals are good and it is never opened. However, in order to serve a useful purpose, components need to be passed into the controlled workspace and finished products passed out. This can be achieved through a number of different specialised transfer devices. An understanding of the principles of operation of these devices is important together with the ability to select and maintain the most appropriate type for your application.

Scope

This chapter considers the various transfer devices with or through which items can be transferred into the isolator and processed components or finished products transferred out for use. It classifies the different types of transfer device in accordance with the classification in the forthcoming standard Draft ISO 14644-Part 7. It recommends transfer devices for various specific applications. Cautionary notes are made about those applications for which some devices are unsuitable. It is recommended that readers be fully acquainted with the different types of transfer device so they can ensure that they have the correct device for their particular application and fully understand its operation.

3.1 Function of different transfer devices

Transfer devices are used for the transfer of materials into and out of isolators and also between isolators, without the transfer of unwanted contamination. Each different application has its own particular requirements. Table 3.1 sets out some applications in the most general terms and shows the objective for the transfer device, the considerations in dealing with airborne contamination, some suggestions as to which classes of transfer device are suitable and the considerations in dealing with surface contamination. Some of these must be taken in combination, e.g. where toxic substances have to be handled aseptically.

Table 3.1 Application of transfer devices for pharmaceutical isolators

Application	*Objective*	*Airborne contamination*	*Device(s)*	*Surface contamination*
Aseptic non-gassed isolator – materials in	Product protection	It should be possible to demonstrate that unfiltered air from the background environment cannot enter the isolator at any time including during transfers	A2 (+ve only) C1 (+ve only) C2 (–ve only) D	There should be a validated spraying and swabbing sanitisation procedure to minimise the risk of contamination entering the isolator on material surfaces
Aseptic non-gassed isolator – materials out	Product protection	It should be possible to demonstrate that unfiltered air from the background environment cannot enter the isolator at any time including during transfers	A2 (+ve only) C1 (+ve only) C2 (–ve only) D	
Non-gassed isolator for hazardous materials – materials in	Operator protection/ containment	It should be possible to demonstrate that unfiltered air from the isolator cannot reach the background environment at any time including during transfers	C2 (–ve only) D	
Gassed and non-gassed isolators for hazardous materials – materials out	Operator protection/ containment	It should be possible to demonstrate that unfiltered air from the isolator cannot reach the background environment at any time including during transfers	C2 (–ve only) D	There should be a validated procedure for the safe removal of product and waste materials that might be carrying toxic contamination
Aseptic gassed isolators – materials in and out	Product protection	It should be possible to demonstrate that unfiltered air from the background environment cannot enter the isolator during gassing, processing and packaging	A1 E F	There should be a validated process for the gaseous sanitisation of the surfaces of materials in the isolator
Aseptic gassed isolators – materials in and out	Product protection	It should be possible to demonstrate that unfiltered air from the background environment cannot enter the isolator at any time	E F	There should be a validated process for the transfer of materials, whose surfaces have been subject to gaseous sanitisation, between gassed isolators
Anaerobic isolators – materials in and out	Product protection	It should be possible to demonstrate that the specified anaerobic conditions exist for all transfers	B2	

Transfer devices should fully achieve their objective of product protection, operator protection (containment), or both, in the background environment in which they are used.

Air-purged transfer devices, namely types C1, C2 and D, should be governed by timed interlocks. It should not be possible to open either door until the transfer chamber is fully purged. In hospital pharmacy practice, timed interlocks should be obligatory, not just to ensure air purging, but also to ensure that the validated contact time with a liquid sanitising agent is achieved.

Separate leak testing of transfer devices is desirable.

3.2 Types of transfer device

The following types of transfer device, which are here represented diagrammatically, are in common use. For any one type of transfer device, it is more important that the rationale for its design is clearly understood than that it complies precisely with its respective description and diagrammatic representation. There is no implication of ranking in this list.

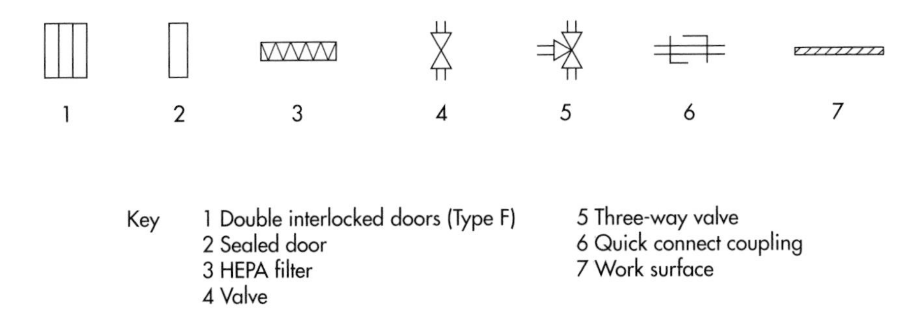

Figure 3.1 Key to transfer device diagrams in Figures 3.2 to 3.10.

3.2.1 Type A1 transfer device

A type A1 transfer device (Figure 3.2) comprises an opening in a separative enclosure which can be sealed by means of a door or similar. There are no means which prevent or minimise the transfer of airborne contamination into or out of the separative enclosure when the door or similar is open. However, when the opening is sealed, the separative enclosure complies with its specified leakage rate.

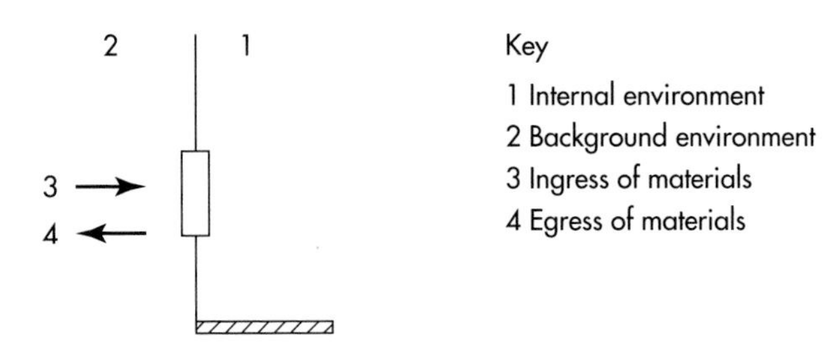

Figure 3.2 Type A1 transfer device.

The main application is for gassed isolators where there is no requirement for materials to enter or leave the separative enclosure during the period that it is sealed. It is not suitable for non-gassed isolators, as every time the door is opened potentially contaminated air from the background environment can enter the isolator.

Examples doors, access panels, zipped panels, 'jam pot' covers, bag-in–bag-out systems.

3.2.2 Type A2 transfer device

A type A2 transfer device (Figure 3.3) comprises a carefully engineered opening in a separative enclosure which relies upon aerodynamic means, namely a controlled flow of air, to prevent or minimise the transfer of airborne contamination from the background environment into the separative enclosure.

The main application is positive pressure enclosures where materials are required to enter or leave continuously.

Examples dynamic holes, mouse holes.

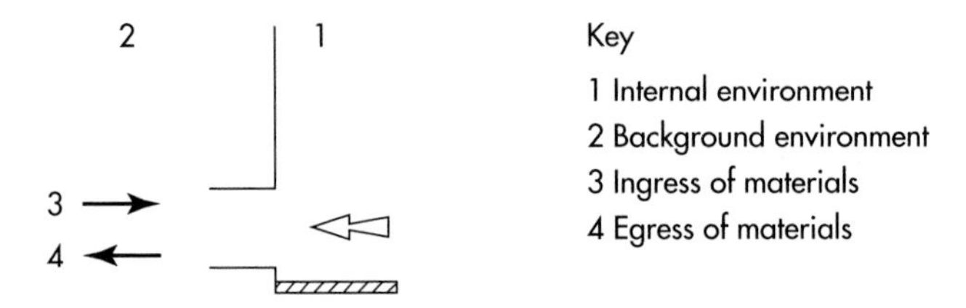

Figure 3.3 Type A2 transfer device.

3.2.3 Type B1 transfer device

A type B1 transfer device (Figure 3.4) comprises an enclosed space attached to the separative enclosure with one sealable opening into the separative enclosure and one sealable opening into the background environment. The direct passage of air between the background environment and the separative enclosure is prevented when either one or both openings are sealed. However, during transfers of material, potentially contaminated air from the background environment can be trapped and then released into the separative enclosure and potentially contaminated air from the separative enclosure can be trapped and released into the background environment.

The main application is where the transfer in or out of potentially contaminated air during transfers of material in or out is either non-critical or can be dealt with by other means.

Examples air-locks, double door sealed transfer chambers, bagging ports.

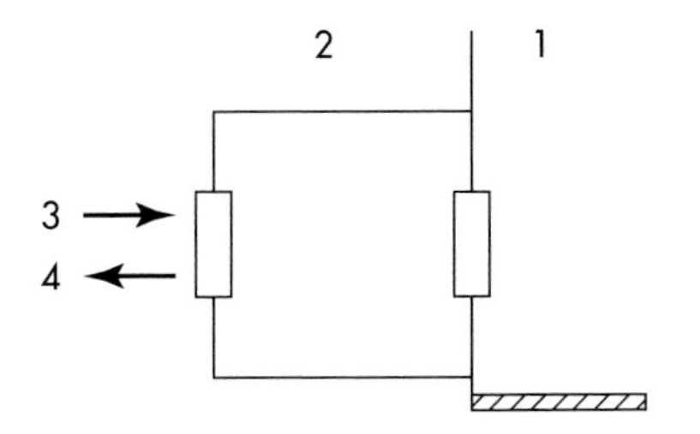

Key
1 Internal environment
2 Background environment
3 Ingress of materials
4 Egress of materials

Figure 3.4 Type B1 transfer device.

3.2.4 Type B2 transfer device

A type B2 transfer device (Figure 3.5) is similar to a type B1 transfer device but has gas and vacuum connections that permit the introduction and total removal of gaseous substances.

The main application of type B2 transfer devices is where gaseous substances are required to be introduced into and/or totally removed from the transfer device. This may be for the purpose of ensuring compatibility between the environment in the transfer device and the environment in the separative enclosure during transfers of material, or

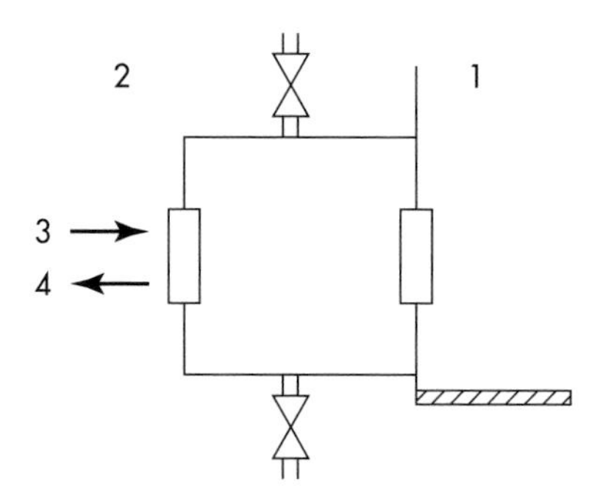

Figure 3.5 Type B2 transfer device.

for the purpose of purging the transfer device where this is required prior to transfers of material in or out.

Example vacuum transfer chamber.

3.2.5 Type C1 transfer device

A type C1 transfer device (Figure 3.6) is applicable to positive pressure separative enclosures only, and comprises a chamber with two doors and a HEPA filter. One door (the 'inner' door) opens into the separative enclosure but is designed to permit the outward passage of air when closed. The other door (the 'outer' door) opens into the background environment. The HEPA filter allows the outward passage of filtered air when the outer door is sealed, thus completing the physical barrier of the separative enclosure.

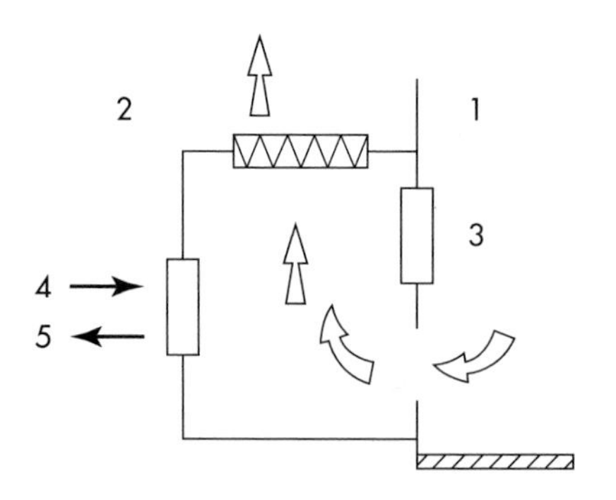

Figure 3.6 Type C1 transfer device.

The main application for type C1 transfer devices is where materials require to be transferred in or out of positive pressure separative enclosures without air from the background environment reaching the internal environment of the separative enclosure. Type C1 transfer devices provide *product protection* for positive pressure separative enclosures during material transfers.

Examples single filtered transfer chambers for positive pressure separative enclosures.

Type C1 transfer devices are not suitable for use with negative pressure separative enclosures because unfiltered air from the background environment would be allowed to reach the internal environment of the separative enclosure when the outer door is open.

Type C1 transfer devices are not suitable where operator protection or containment is required in positive pressure separative environments because unfiltered air from the internal environment would be allowed to reach the background environment when the outer door is open.

3.2.6 Type C2 transfer devices

A type C2 transfer device (Figure 3.7) is a particular design that is applicable to negative pressure separative enclosures. It comprises a chamber with two doors and a HEPA filter. One door (the 'inner' door) opens into the separative enclosure. The other door (the 'outer' door) opens into the background environment. Both doors seal. The HEPA filter completes the physical barrier of the separative enclosure when the outer door is sealed whilst allowing the inward passage of filtered air into the (transfer) chamber. Air from the (transfer) chamber (which may be filtered if the outer door is closed or unfiltered if the outer door is open)

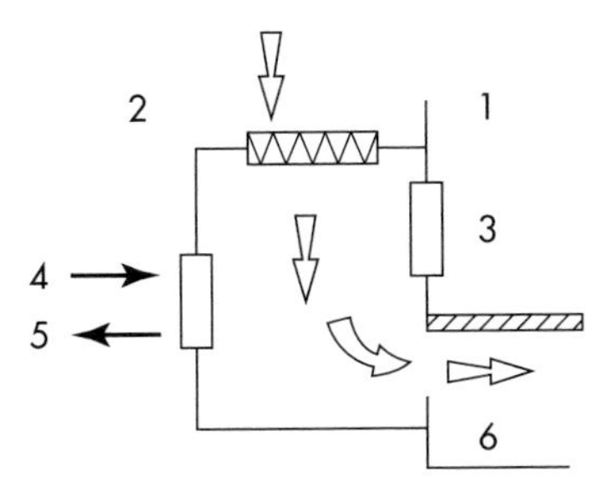

Key
1 Internal environment
2 Background environment
3 Negative pressure
4 Ingress of materials
5 Egress of materials
6 Exhaust

Figure 3.7 Type C2 transfer device.

is drawn through a slot into the space underneath the work surface of the separative enclosure and on into its exhaust system. In some designs, the internal environment of the separative enclosure and the space beneath its work surface are connected by airslots that form part of the air path in the separative enclosure. In this case, the pressure regime in all operational situations must be such as to prevent any airflow back into the internal environment of the separative enclosure.

The main application for type C2 transfer devices is where materials require to be transferred into or out of negative pressure separative enclosures without any air from the background environment reaching the internal environment of the separative enclosure or unfiltered air from the the separative enclosure reaching the background environment when the separative enclosure is in its operating state. C2 transfer devices provide *product protection* and *operator protection (containment)* for negative pressure separative enclosures during material transfers.

Example single filtered transfer chambers for negative pressure separative enclosures.

Where operator protection (containment) is required, type C2 transfer devices are not suitable for use with positive pressure separative enclosures because unfiltered air from the internal environment would be allowed to reach the background environment of the separative enclosure when the outer door is open.

3.2.7 Type D transfer devices

A type D transfer device (Figure 3.8) comprises a chamber with two doors and two HEPA filters. One door (the 'inner' door) opens into the

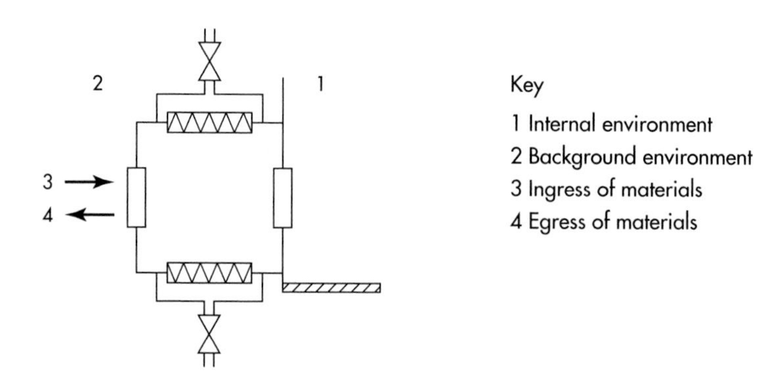

Figure 3.8 Type D transfer device.

separative enclosure. The other door (the 'outer' door) opens into the background environment. Both doors seal. Air enters the (transfer) chamber through one of the HEPA filters and leaves through the other. In some designs, type D transfer devices may form part of the filtered airflow system of the separative enclosure when one of the HEPA filters is placed between the (transfer) chamber and the internal environment of the separative enclosure. They may be used with both positive and negative separative enclosures and are used to provide *product protection* and *operator protection (or containment)* during material transfers.

Examples double filtered transfer chambers, double filtered separative enclosures used for transfers.

3.2.8 Type E transfer devices

A type E transfer device (Figure 3.9) is a type D transfer device that is fitted with additional gas inlet and outlet connections. These connections may be upstream of the inlet HEPA filter and downstream of the outlet HEPA filter, or direct to the inside of the (transfer) chamber. A type E device is used where there is a requirement for a gaseous substance or vapour to be introduced into the transfer device as part of the process being carried out in the whole separative enclosure system.

Examples gassable/autoclavable transfer devices including certain transfer separative enclosures and docking devices, permanently connected autoclaves, sterilising tunnels and similar devices.

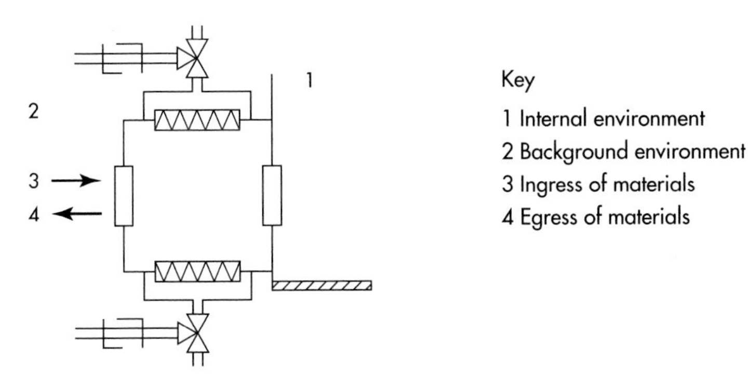

Figure 3.9 Type E transfer device.

3.2.9 Type F transfer devices

Type F transfer devices (Figure 3.10) describe the family of 'double-door' transfer ports which, by careful mechanical design and manufacture, can be used to make immediate transfers without break of containment. They are commonly known as rapid transfer ports (RTPs) but are also referred to by manufacturer's names such as 'DPTE' (double-porte de transfert étanche) or 'HCT' (high containment transfer). They may also be referred to as 'docking ports'. Type F transfer devices provide *product protection* and *operator protection (containment)* during transfer of materials between different isolators or other sterilising devices that are not connected to each other.

The type F transfer device consists of four major components:

1. Port flange, also known as the alpha flange or the female flange. This is normally mounted rigidly on the wall of an isolator.
2. Port door. This closes into the port flange and carries a special 'arrowhead' seal.
3. Container flange, also known as the beta flange or male flange. This is normally fitted to the front of a mobile container and carries a further 'arrowhead' seal.
4. Container lid. This closes the container.

The operation of the device is shown in Figure 3.11.

The design of the device brings the two arrowhead seals together point-to-point and thus minimises, within engineering limits, the common potentially contaminated area around each seal. This narrow ring of potential contamination may be small but nevertheless represents a risk. It is therefore often termed the 'ring of risk'. Some of the more complex devices of this type attempt to reduce the risk by the use of local heat or bright visible light. Risk may also be minimised by simple operational techniques such as the manual application of a sanitising agent to the

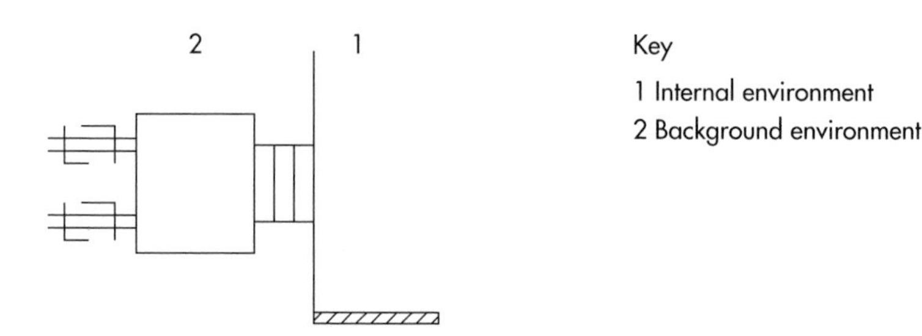

Figure 3.10 Type F transfer device.

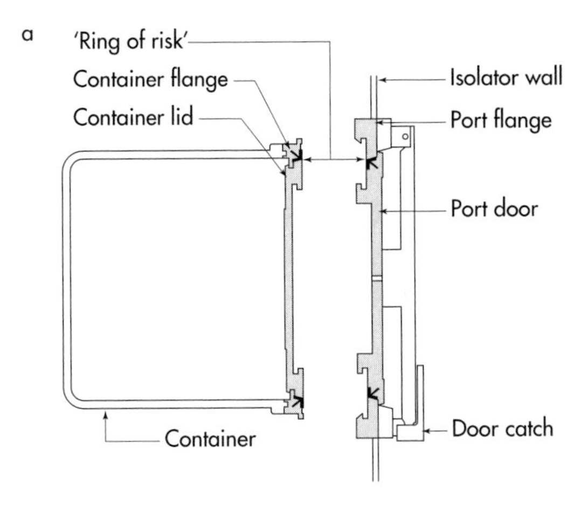

a. The container, complete with container lid, is 'docked' onto the port flange, whilst the port door is closed. This process may be a bayonet action, the container being presented to the port flange by the operator and rotated approximately 60 degrees onto stop pins. Alternatively it may be a clamping action in which case there is no rubbing of the arrow-head seals.

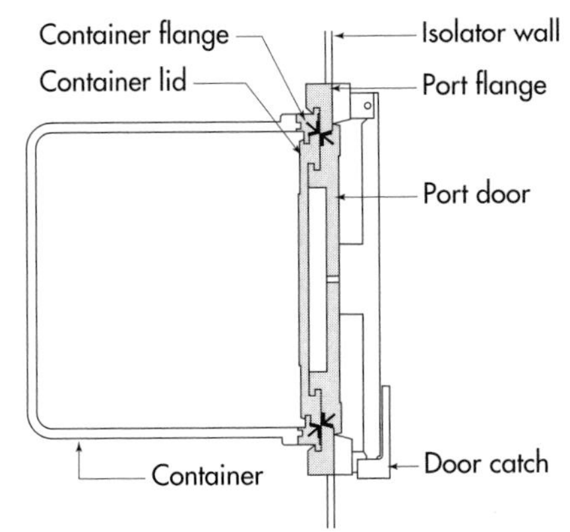

b. The docking process sequentially locks the container flange to the port flange, releases the container lid from the container, and locks the container lid to the port door.

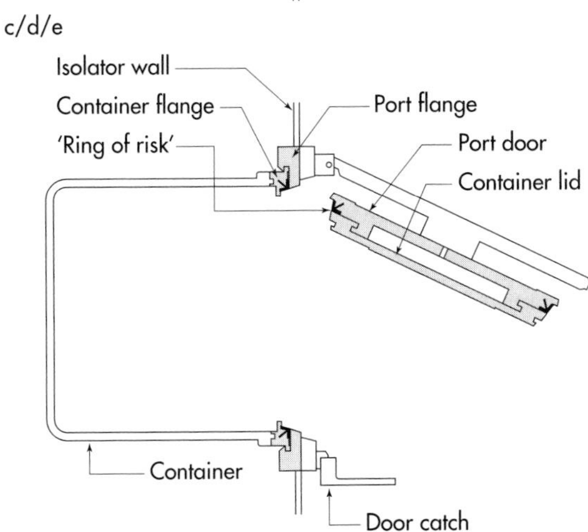

c. The operator, now working from inside the isolator, opens the port door, this being locked to the container lid to form the 'double-door'.

d. Items may now be transferred from container to isolator or vice-versa with no transfer of contamination from the background environment to either the container or the isolator and no transfer of containment from the container or isolator to the background environment.

e. Following transfer, the port door is closed and the container may be 'undocked' from the port and removed.

Figure 3.11 Operation of the type F transfer device.

seals immediately prior to docking and protection of the seal area by covering during material movement.

Containers may take a number of forms, but each must be sanitised prior to use. Simple plastic drums may be sanitised internally by attachment to the isolator during gassing or they may be fitted with HEPA filtered connections and sanitised internally by direct gassing for rapid processing. Stainless steel containers may be fitted with hydrophobic filters and autoclaved, complete with their contents, as may Tyvek containers. Alternatively, the container may comprise a complete mobile isolator by fitting it with a container flange on a suitable flexible connection. In this way, a transfer isolator may be used to move quantities of materials around from isolator to isolator, without lengthy sanitisation stages.

Some designs will include a mechanical interlock which prevents the opening of the port lid if a container is not present, and this is a significant advantage. Other types use cheap single-use disposable containers, removing the need for cleaning and sanitisation. Yet another design is the 'split butterfly' valve such as the Buck port. This is especially useful in bulk powder or granule transfer for heavier industrial applications, both for sterile and toxic containment transfers. Users should consult with the manufacturers to decide which type of Class F transfer device will best suit their application.

4

Access devices

Gloves and gauntlets are the most common way of gaining access to the materials to be manipulated inside the isolator. They are usually the most vulnerable components, prone to wear and leakage, and so require special consideration. Alternative means of access such as robots or other mechanical manipulative systems, suggest the use of a generic term, hence the phrase 'access devices' is used.

Introduction

Access device is the generic term used to describe the device or means of access used from outside an isolator to carry out manipulative work inside an isolator. It is specifically designed to maintain the separation or barrier.

Access devices can range from glove/sleeve systems to robots. By contrast, a device which allows the transfer of materials between the inside and the outside of the isolator is classed as a transfer device and is dealt with in Chapter 3: Transfer devices.

Access devices should be compatible with the work to be carried out. Ergonomics and material compatibility are the principal factors in selecting the appropriate device. The position and mounting are largely determined by ergonomics.

4.1 Gloves

4.1.1 General requirements

The simplest and most common access devices are gloves which are fitted to glove/sleeve systems. Gloves are the most critical component of an isolator because:

- they contain items which are potentially highly contaminated – the operator's hands;
- they contain items which are susceptible to toxic materials – the operator's hands;

- they are likely to be in contact with the work and with surfaces in the isolator. At best they are in close proximity;
- leaks may be difficult to detect.

At the same time, gloves are the most vulnerable component of the isolator envelope or barrier for the following reasons:

- they are made of thin materials;
- they are prone to wear and tear;
- they are prone to physical damage;
- they are prone to chemical damage;
- they are prone to damage as they are changed, which may be at frequent intervals.

Important considerations for gloves are:

- They should be made of thin materials for operator touch sensitivity.
- They should be made of a material that is sufficiently flexible for operator dexterity.
- They should be made of a material that is sufficiently elastic for the glove change system.
- They should be free of leaks.
- They should not be capable of contaminating the product in aseptic applications. This predicates that they should be powder-free and sterile.
- They should provide the required level of operator protection subject to a risk analysis. For toxic substances, this means that they should have sufficient impermeability. For radiopharmaceuticals this means that they should provide sufficient radiation protection.
- They should be resistant to any chemicals or cleaning agents likely to be used, including, where applicable, gaseous sanitising agents.
- They should be capable of being cleaned.
- They should be made of a material which has sufficient strength and durability.
- They should, preferably, be made of non-allergenic material.
- If required, they should be made of a material that has suitable electrostatic properties.
- Where double gloving is used, i.e. where the operator is already wearing a glove before entering the isolator glove, the materials of the respective gloves should not bind or adhere to each other.
- They should be the correct size for the operator, and a range of sizes should be accommodated by the glove change system.
- They should be suitable for the glove change system. Usually this means they should be beaded.

- They should be as comfortable to wear as possible.
- They should be cost effective. Gloves that are changed frequently should cost considerably less than gloves that are sufficiently durable for long-term use.
- They should carry a CE Mark and, if particular requirements are specified, suitable certification should be available.

Compromises will need to be made, for example operator dexterity should be balanced against glove strength. Testing is covered in Chapter 8: Leak testing. It should be noted that cleanroom or isolator glove properties are not necessarily the same as those for gloves used directly in medical or surgical procedures.

4.1.2 Glove type

Gloves may be handed or ambidextrous. Glove-change systems usually require gloves to have a bead ring at the wrist. Such gloves are known as beaded gloves. The length of the wrist section is another variable.

4.1.3 Glove material

Gloves are available in a wide variety of materials and Table 4.1 provides some information on the most commonly used materials.

Information about some of these and other materials is given in Draft ISO 14644-Part 7.

Manufacturers should be expected to supply test data on their gloves, but additional testing or evaluation may be performed by the user to determine the most suitable material for a particular process application.

4.1.4 Glove sizes

Gloves are still manufactured in accordance with national codes and standards. These may vary from country to country in respect of thickness, length, diameters and other features such as finger dimensions. International standards are now being established and manufacturers are moving towards common dimensions and features.

In Europe, gloves generally comply with BS EN 420 and are produced either as anatomical, i.e. different left and right hand or as ambidextrous, i.e. suitable for either hand.

It is worth noting that gloves produced in the USA tend to have slightly shorter fingers than those produced in continental Europe. Where

Table 4.1 Most commonly used materials in gloves

Material	*Advantages*	*Disadvantages*
Natural rubber latex	Exhibits a softness and feel not yet achieved with other synthetic polymers. Resists most moderate chemicals, organic acids, alcohols, ketones and aldehydes. Is relatively inexpensive.	Attacked by ozone, strong acids, fats, most hydrocarbons, and degrades with sterilising processes. In particular is fairly susceptible to hydrogen peroxide vapour. Can provoke an allergic reaction.
Neoprene	A general-purpose elastomer, resists moderate chemicals, acids, ozone, oils and solvents. Has a high tensile strength and flexibility and is good for sensitivity and dexterity.	Attacked by strong oxidising acids, esters, chlorinated aromatics and nitro-hydrocarbons.
Nitrile	Has good solvent resistance, is very durable and has good puncture resistance.	Resists hydrocarbons and many chemicals, is attacked by ozone, ketones, esters, chlorinated and nitro-hydrocarbons. Can be less elastic.
'Hypalon'®	Has good resistance to oxidising chemicals, ozone, many acids, vapour hydrogen peroxide sterilisation, heat and abrasion. Very durable, has a high tensile strength can be produced in varying thicknesses down to 0.008 in (0.02 mm).	Is attacked by concentrated acids, esters, chlorinated and nitro-hydrocarbons. Relatively expensive.
Butyl	Has a very low permeation of water vapours, gases and toxic chemicals, is resistant to oxygenating solvents and most oxidising chemicals and is flexible at low temperatures.	Is attacked by petroleum solvents, coal tar solvents and aromatic hydrocarbons.
'Viton'®	Has excellent resistance to permeation from many chlorinated, aliphatic and aromatic hydrocarbons, most oils and acids. Is good for dexterity.	Is attacked by ketones, low molecular weight esters and nitro containing compounds.
EPDM (Ethylene propylene diene monomer)	Is soft and has good tensile strength, ozone and vegetable oil resistance. Used for high pressure and aggressive environments.	Is attacked by mineral oils and solvents.
Polyurethane	Thin film – excellent sensitivity, dexterity and tactility. Latex-free so no latex protein allergies. No process chemicals so virtually eliminates dermatitis allergies. Micro-roughened surface – excellent grip even when wet. Low particle shedding. Low level of extractables. Excellent ESD properties so reduces ESD related effects. High tensile strength. Beaded cuff. Resistant to a wide range of chemicals.	Still under evaluation

the dexterity of the operator is important, it is desirable that the fingers of the operator reach the end of the glove without any slack. Factors that govern glove size selection are:

- physical characteristics of the operator/s;
- male or female operator/s;
- loose fit or form fitting;
- use of glove liners;
- use of double gloving;
- need for ventilation;
- operator comfort.

4.1.5 Glove manufacture

Most gloves are produced by dipping hand-shaped formers into a liquefied material. Where the liquefied material is a water-based emulsion, the hand-shaped former is first dipped into a coagulant, and then into the emulsion. Where the liquefied material is a solution, the hand-shaped former does not need any preparation. Gloves produced by the latter process are less likely to shed. Latex, neoprene and nitrile gloves are produced by the emulsion dipping process. Hypalon®, butyl, Viton® and EPDM (ethylene propylene diene monomer) gloves are produced by the solution dipping process. The size of the glove is determined by the size of the former. The thickness of the glove may be increased either by repeated dipping or by longer dwell times in the rubber solution until the required thickness is achieved.

4.1.6 Variable thickness gloves

Gloves can be produced with a dual thickness, where the thickness varies around the perimeter with a thinner part on the finger or palm zones, to provide enhanced touch sensitivity.

4.1.7 Laminated and loaded gloves

Gloves can be produced in a laminated combination of two or even more materials to take advantage of their different properties. An example is where the flexibility of neoprene is combined with the ozone and chemical protection of Hypalon®. Loading is where an additional ingredient is incorporated into the raw material. An example of this is ‘lead-loading’ to provide radiation protection.

4.1.8 Glove properties

4.1.8.1 Allergic reactions

Natural latex rubbers may be allergenic. This is because they contain proteins that are capable of producing an allergic reaction in some people. This effect is related to the natural defence mechanisms of the rubber tree and is best dealt with by either:

- specifying good quality gloves with a low protein content;

  Note: EN 455-Part 3:2000 Medical gloves for single use, defines a method for determination of protein levels. This is the modified Lowry assay test which is accurate to approximately 10 ppm protein per gram of glove material. Research has suggested that levels of less than 50 ppm should not cause problems, even to those already sensitive to latex, but may cause problems to the very small percentage of the population considered to be 'hypersensitive' to latex.

- wearing inner gloves of a different material.

4.1.8.2 Quality and leaktightness

It is of prime importance to use gloves that do not leak. It has been alleged that the best quality of latex is reserved for another very critical product, namely condoms. Users should be satisfied about the quality and leaktightness of the products they are using. More information may be available from specialist suppliers or from independent testing laboratories. See also Chapter 8: Leak testing

4.1.8.3 Non-shedding

Gloves to be used in the controlled workspace of an isolator for aseptic work should be manufactured from a material that does not shed particles. It should be noted that latex, neoprene and nitrile gloves, which are made by the emulsion dipping process, may shed particles if they are not wiped, for example with IPA, before each work session.

4.1.8.4 Powder-free

Gloves to be used in the controlled workplace of an isolator for aseptic work should be supplied without lubricating powders. This eliminates one source of particulate contamination for the product or controlled workspace.

4.1.8.5 Sterility

Gloves for use in isolators for aseptic processing should normally be supplied in a sterile state and suitably packaged. Alternatively, gloves may be treated *in situ* as part of the sanitisation or sporicidal gassing process of the isolator.

4.1.8.6 Resistance to cytotoxics

Studies have shown that cytotoxic solutions pass through virtually all glove materials given time. For this reason, gloves should be changed regularly when used for manipulating cytotoxic preparations and consideration should be given to double-gloving.

4.1.8.7 Packaging and storage

Packaging should be appropriate to protect the glove and ensure it is in a suitable state and ready for use with the isolator.

Storage is particularly important. All elastomers degrade over time and the aging characteristics of specific polymers are determined by the anti-degradation additives present and by the chemical structure of the polymer.

Therefore, the supplier's method of packing and storage should be considered when selecting gloves. Shelf-life and 'in-house' stock rotation procedures are important control factors for the user.

4.1.9 CE marking and standards

All gloves used within the European community should bear a 'CE' Mark. This signifies that they conform to the basic health and safety requirements as set out in Personal Protective Equipment Directive 89/686/EEC.

Depending on application, gloves should be tested against the appropriate European standards. These may be one or more of the following shown in Table 4.2.

4.2 Sleeves

4.2.1 General properties

Gloves are usually used with sleeves which are sealed to shoulder rings set in a window or visor, or, in the case of flexible film isolators, the

Table 4.2 European Standards for gloves

Standard	*Title*	*Performance test*
EN 388	Gloves against mechanical hazards	Abrasion, blade cut, tear, puncture resistance
EN 374 Part 2	Gloves against chemical and microorganisms	Air/water leak test methods (batch test acceptance levels)
EN 374 Part 3	Gloves against chemical and microorganisms	Chemical permeation (test methods give a breakthrough time in minutes for a chemical against a given glove material)
EN 421	Gloves against ionising radiation and radioactive contamination	Water vapour permeability Ozone resistance Attenuation efficiency (lead equivalence mm)

canopy. Occasionally the sleeves are welded to the canopy so they form part of it. The sleeves are fitted with cuff rings to carry the gloves, which may be changed by various means including proprietary devices. The logic here is that the gloves are the most vulnerable part of the isolator system and should be easily changeable, preferably without break of containment. The second most vulnerable parts are the sleeves, and these too should be easily changeable, though not necessarily without break of containment.

The material of sleeves should be tough enough to withstand continual folding and stretching, but flexible and supple, so that they do not restrict the user. The surface exposed to the isolator should be smooth and easily cleaned, whilst the operator side should be pleasant to the touch and neither sticky nor abrasive.

Some types of sleeve, made up from two separate non-bonded layers of material, are designed so that if either layer becomes punctured, then the two layers separate visibly, making leaks in the sleeves self-indicating.

4.2.2 Accordion sleeves

Sleeves can be constructed in an accordion or corrugated format to enhance movement and extension. This concept may be considered for some pharmaceutical isolators. Its main features are:

- prevention of reversing outwards in positive pressure environments;

- enhancement of operator comfort by holding the sleeve off the operator's arm and permitting an airflow.

However, an accordion sleeve is not easy to clean and thus may not be suitable for aseptic applications.

4.2.3 Cuff rings and glove changes

Cuff rings are normally of moulded or machined plastic, with grooves to retain the beaded gloves and sleeves and any security o-rings. They should be as light and unobtrusive as possible, but strong enough to withstand the glove change process.

4.2.4 Changing gloves

The simplest cuff rings only retain the sleeve and glove but do not permit glove change without breaking the barrier.

Slightly more complex rings allow aseptic glove changes by the glove-over-glove method to maintain the barrier. In this case, the bead of the glove to be changed is first moved to the forward groove. Then a new glove is placed entirely over the old one, with its bead in the rearward groove. With the new glove in place the bead of the old glove is eased out of its place and into the sleeve for removal. The isolator manufacturer may provide special sleeves, with widened wrists to allow for the change process.

Where hazardous substances such as cytotoxics are present, a safer variation of the aseptic glove change may be used. Here the operator should be wearing a protective glove before starting the glove change. The glove to be changed is inverted into the sleeve and held before removal. An impregnated wipe is used to clean the contaminated surface of the old glove around the cuff area and is then inserted into the inverted old glove for disposal. The new glove is then placed over the old glove and secured and the old glove removed. Finally the operator removes his own protective glove, turning it inside out to enclose the old contaminated glove and impregnated wipe for safe disposal.

Another type of cuff ring has a telescopic action, with the sleeve retained on an outer ring and the glove retained on an inner ring. Glove changes can be made by displacing the old inner ring and glove and pushing a new inner ring and glove into place. This system maintains the barrier and allows glove changes in either aseptic or containment isolator mode depending on the direction from which the new inner ring with glove is introduced.

4.2.5 Leak testing

Since the gauntlet or glove/sleeve assembly is a critical part of the system, isolator manufacturers often supply a device for leak testing the assembly independently from the body of the isolator. Such a device usually consists of a disc, which seals onto the shoulder ring, and a micromanometer to monitor pressure decay. The initial pressure differential may be the existing pressure differential between the inside and the outside of the isolator, or may be increased by inflating the sleeve. Glove sleeve testing is discussed in Chapter 8: Leak testing.

4.2.6 Shoulder rings

As with cuff rings, these are normally of moulded or machined plastic. In the case of rigid isolators, they are usually sealed into the isolator window, although occasionally they are mounted in the isolator wall. In flexible film isolators, they are fixed and sealed onto the canopy.

For negative isolators, the shoulder ring is mounted facing out of the isolator and the sleeve or gauntlet is changed from outside the isolator.

For positive pressure isolators, the shoulder ring is mounted facing into the isolator and the sleeve or gauntlet is changed from within the isolator. Because it is difficult to change sleeves in this configuration, they may be mounted outside as for negative pressure isolators, even for aseptic operations. This may inhibit penetration of sanitising agents and especially sanitising gas, to the area of the sleeve close to the shoulder ring.

Where shoulder rings are fitted to form access ports for infrequent use, beaded port seals or 'blanking caps' are available which avoid the need to leave gauntlets in the potentially degrading environment of the isolator. Contained changes are possible (see section 4.3).

4.2.7 Provision for gassing

In applications where gas sanitisation is used, there should be provision for sleeves and gloves to be supported in an extended position during sanitisation. Suitable devices include inflatable inserts and floor-standing supports.

4.3 Gauntlets

A gauntlet is effectively a one-piece sleeve and glove combined. It may be full or mid-arm length and the length is defined accordingly.

- Gauntlets are particularly suitable for hazardous containment situations and can be produced in many materials including laminated materials and radiation protecting materials.
- Gauntlets are produced in various sizes and configurations to the same European standards as gloves. As with gloves, they may be shaped to be anatomical or ambidextrous.
- Gauntlets are not as sensitive in use as gloves and are generally used for more arduous or intensive operations where frequent changes are not appropriate. They are often used by more than one operator. In such situations, for hygiene, it is recommended that operators don suitable 'liner' gloves before entering the gauntlets.

The frequency of changing for gauntlets is normally extended to the maximum practical period. This is due to their high cost and the difficulty of carrying out a change without breaking the barrier. Gauntlets for use with isolators are thicker and more durable than gloves and intended to be used for longer periods. A gauntlet has a defined operational life and should be changed well before this is reached.

It is not easy to carry out gauntlet changes manually without breaking containment. There are also several proprietary mechanical designs that facilitate changes in toxic applications, but these are all essentially heavy nuclear systems that have not found use in pharmaceutical applications.

4.4 Half-suits

Gauntlets and glove/sleeve assemblies give a working radius of about 500 mm within the isolator and a normal lifting capacity of up to 5 kg. Where greater reach and lifting ability is required, the isolator can be fitted with a half-suit. This covers the user from the waist upwards and extends the reach to around 1200 mm with a lifting capacity of perhaps 15 kg. Operators usually work from a standing position, but a high chair can be used to allow comfortable seated operation.

A half-suit is normally mounted on the isolator base-tray or floor and may be on a raised or angled plinth to make entry and exit easier. A half-suit is fitted with a clear vision helmet which may be a separate, changeable item. It is also fitted with cuff rings on its sleeves to carry gloves. Half-suits need to be light, yet tough and flexible. A platform may be provided underneath the isolator so that operators of shorter stature can work comfortably. This can take the form of a fold-down platform or an adjustable powered platform.

All half-suits have a supply of breathing air which is fed to the neck region and also the wrist region. Some suits have a double-layer structure in which the inter-space forms the air duct. Others have air ducts formed in the flexible film over the surface of the suit.

Breathing air is most often supplied by a centrifugal fan. Air from the background environment may pass though a HEPA filter for added safety and comfort of the operator.

Operators should **never** enter a half suit when the air supply is switched off. Should the supply fail at any time, the operator should leave the suit immediately.

Operators are advised not to enter the suit during gas sanitisation of the isolator. After gassing of the isolator is complete, the air supply to the suit should be run for an approved minimum time, probably at least 30 min, before the operator enters. This is because the sanitising gas can reach relatively high concentrations in a half-suit by diffusion, depending on the gassing time.

Half-suits should be supplied with means of support when unoccupied to prevent collapse out of the isolator and to expose all surfaces during gas sanitisation. Such devices should be simple to operate and should not encumber the operator.

Entering a half-suit whilst the isolator is under positive pressure is not entirely easy and consideration should be given to suits with means to aid entry, such as inflatable support rings to hold the half-suit off the operator. This is especially relevant in a positive pressure isolator where the isolator pressure presses the half-suit onto the operator.

Communication from half-suits is not easy and consideration should be given to the provision of intercom systems.

Half-suits should be introduced to operators carefully to avoid adverse reaction such as claustrophobia. Compared with open-bench operation, half-suits are encumbering devices, but some operators prefer them to glove/sleeve systems.

4.5 Full suits

In some large-scale pharmaceutical processes, full man-entry may be required and at least one proprietary design of suit exists to allow this. The suit is based on the type F transfer device as described in Chapter 3: Transfer devices. The double-door principle is used in conjunction with a two-part suit to allow the operator to be 'posted' into the enclosure without break of containment. Umbilical lines supply breathing air, return exhaust air, and provide for communication. The original design was for nuclear work and the full suit has seen little application so far in the pharmaceutical industry.

4.6 Robotics

Some isolator work, such as routine sterility testing on large batches of product, is very repetitive. There has been some success with the use of small industrial robots in isolators for this type of work and it seems likely that this may be expanded in the future. The complex structure of the robot can be covered with a flexible gaiter similar to those used on master–slave manipulators in the nuclear industry, to provide for gas sanitisation of the enclosure.

In robotics, as with any other machinery inside an isolator, careful consideration should be given to safety interlocking to prevent operator access when moving parts are present. Light beam arrays have been used successfully for this.

The following points should be considered when designing robotics for use in isolators:

- Does the equipment compromise the internal environment with regard to microbiological or particulate contamination?
- Will a flexible cover or gaiter be required to protect the internal environment from particles, grease and oil?
- Does the equipment disturb designed air movement within the isolator?
- Will smoke tests be required?
- If unidirectional airflow is used, is the airflow still unidirectional in the critical zone(s)?
- Are there standing vortices in the critical zone(s)?
- Are adequate clean-up rates achieved?
- Does the equipment, or electrical/pneumatic connections to the equipment, compromise the integrity of the isolator? Leak paths can be a problem and such paths exist, even through cable cores.
- Can the equipment, including any protective gaiters, carry out its full range of movement without affecting other equipment or access devices, e.g. gloves within the isolator?
- Can the equipment be easily accessed for adjustment or maintenance without breaking containment?
- Does the equipment and operating system conform to GAMP?
- What is the reliability of the system?
- In the event of a failure can product be recovered or would batches

in process need to be scrapped? For recovered product, how can product quality be verified?
- Is there a back-up system or alternative method for maintaining essential production in the event of equipment failure?

Finally, the human factors should not be overlooked:

- Do staff have the necessary expertise to operate the system?
- What recruitment and training may be required?
- Is there sufficient local expertise and adequate training to be able to carry out basic fault finding, maintenance and recovery?
- What back-up maintenance contracts will be required to ensure adequate response and reliability?
- Is there access to personnel or expertise to help specify and validate the project properly, particularly if the system is to be designed to GAMP? (The volume of paperwork for a GAMP project can be substantial.)
- What safety systems are required to prevent personal injury and equipment damage?
- Have HAZOP studies been carried out?
- Are existing management and staff committed to the new equipment?

4.7 Training and user maintenance

Training is a vital factor in any pharmaceutical operation and training in the use of access devices in isolators is no exception. Training in the use of gloves, gauntlets and half-suits is addressed in Appendix 2: Training.

Note that, as a general precaution, gloves and gauntlets should be changed regularly as a matter of course and not left until they fail.

The time between changes will depend very much on the nature of the work and the type of glove or gauntlet. Experience will determine a safe service period, but Table 4.3 gives a very approximate guide.

Where sleeves and suits are found to have failed, repairs may be made as an interim measure. Some manufacturers provide a ‘repair kit’. Any repairs made to these access devices should be tested for leakage as described in Chapter 8: Leak testing.

Consideration should be given to cleaning and re-sanitisation of aseptic isolators following such repairs.

Table 4.3 An approximate guide to the safe service period for gloves and gauntlets

Glove type/material	*Change interval*
Thin latex (e.g. surgical gloves)	Every work session
Medium latex (e.g. domestic gloves)	Daily/weekly
Industrial nitrile or neoprene gloves and gauntlets	Weekly/fortnightly
Hypalon® gloves and gauntlets	Monthly/3 monthly
Heavy gauntlets	Monthly/3 monthly

5

Siting of isolators and clothing regimes

There has been a range of views on just where is an acceptable place to put an isolator and whether any particular grade or standard of clothing should be worn. Some places are acceptable and other places are not. In just the same way, some standards of clothing are suitable and some are not.

Scope

This chapter describes the environment in which an isolator should be sited and operated. It considers construction, access, environment, support rooms, clothing and product segregation. Existing facilities may not comply with all these requirements. This does not necessarily indicate that such facilities are unacceptable, but deficiencies identified should be critically assessed.

Much of the guidance in this chapter, including the specifications for grades of air and clothing requirements, is taken from Annex 1 of EC GMP.

5.1 Construction

Isolators used for aseptic preparations in hospitals should be sited in a dedicated room, used only for the isolator(s), ancillary equipment and related activities.

Isolator rooms and isolator process areas should not be susceptible to contamination from services or equipment located close by, especially drains.

The rooms should be constructed in such a way that the internal surfaces, walls, floor and ceilings, are smooth, impervious and free from cracks and open joints. The materials used should be non-shedding, sealed and resistant to repeated treatment with cleaning agents and disinfectants.

There should be a minimum of ledges. Shelves, cupboards and equipment in the room should be easily cleanable and free from dirt traps. There should be no uncleanable areas such as crevices or inaccessible recesses.

False ceilings with lay-in tiles are not favoured and should be replaced with a full membrane. If for any reason this is not possible, then each tile should be of a non-shedding material and must be carefully sealed into the supporting grid with a non-hardening sealant such as RTV silicone.

Sinks and other hand washing facilities must not be located in isolator rooms. Hand washing facilities are normally provided at the entry to the change room. Where space is restricted they may be located inside the change room on the dirty side. The hand washing facilities should be controlled with an appropriate disinfection regime and microbiologically monitored.

5.2 Access

The isolator room should be entered through a change room. The doors should be fully interlocked, or fitted with a stop/go style of audible and visual alarm, to prevent both doors being opened simultaneously. The change room should be flushed with HEPA filtered air. Demarcation lines, or preferably step-over barriers, should be used to separate the different stages of change. If any of these features are not present, appropriate disciplines should be included in the SOPs.

5.3 Isolator room

For aseptic applications, the EC GMP recommends at least Grade D air classification for the isolator room. In some cases this is improved to Grade C or even Grade B through the use of terminal HEPA filtration. Pressures should cascade down from here to lower grade areas. EC GMP recommends 10–15 Pa pressure difference between adjacent rooms of different grades. The pressure drop between rooms should be indicated and values recorded at defined intervals. The pressure drop across at least one filter in each room should be indicated. Values should be recorded at defined intervals.

Environments are graded by the EC GMP according to measured levels of particulate contamination 'at rest' and 'in operation' with recommended limits for microbial contamination 'in operation'. The room and the air change rate should be designed to enable the specified 'at rest' conditions to be achieved after a short 'clean-up period' of 15 to 20 min after completion of operations. To achieve this clean-up time a minimum of 20 room air changes per hour is recommended. The clean-up time should be validated. It is of little value to check the 'at rest' state when the room has been allowed to settle for a long period, e.g. first thing in the morning.

The room ventilation system should be provided with a latched alarm indicating power failure.

The level of contamination is influenced by the room layout, the work being carried out, the number of operators, the type of clothing worn, materials being handled (including packaging) and the nature of the equipment.

If containment of hazardous materials is required for pathogenic, highly toxic, radioactive, viral or bacterial materials, then negative pressure rooms are sometimes specified. In such cases, additional measures should be taken to ensure an aseptic environment is maintained. Advice should be obtained from specialist cleanroom designers.

The isolator room should not be used for storage. Only essential items for immediate use should be present. This also applies inside the isolator transfer devices and work zone.

5.4 Support room

The support room is provided so that activities in the isolator room can be limited to the essentials and to reduce the viable and non-viable particulates on materials entering the isolator room. The air classification should be to Grade D if practicable. Activities in the support room include:

- storage of raw materials and components;
- generation of documentation;
- component assembly;
- initial surface decontamination of materials prior to transfer to the isolator room;
- labelling, checking and packaging of completed product for onward despatch.

Materials can be passed into and out of the isolator room through double door transfer hatches, the doors of which should be interlocked. To optimise workflow and ensure materials segregation, departments with a high workload should use separate dedicated transfer hatches for passing materials into and out of the isolator room. Access to the support room should be restricted to authorised personnel.

5.5 Clothing

Appropriate clean room clothing should be worn by personnel to reduce the amount of contamination in the environment. Clothing regimes as defined in Annex 1 of the EC GMP are given in Table 5.1.

For isolator rooms, the minimum standard of clothing should be appropriate for Grade D. It is recommended that the material of the clothing should be non-shedding, that sleeves be gathered at the wrists and that gloves be worn. Typically Grade D clothing is re-useable and does not require specialised laundering. Clothing for Grade B may be either disposable or re-useable. Re-useable Grade B clothing requires specialist laundering and sterilisation.

Where terminal filtration has improved the room air grade to Grade B or C the use of D grade clothing is still acceptable when working with isolators. By comparison, where a Grade B room is the background environment for an open-fronted laminar flow cabinet, then the more stringent clothing regime for Grade B rooms outlined in EC GMP must be observed.

The same clothing regimes apply in isolator rooms where cytotoxics are handled. In the event of clothing being accidentally exposed to cytotoxics, it is recommended that it should be disposed of as contaminated waste.

Table 5.1 Clothing regimes as defined in Annex 1 of the EC GMP

Room grade	*Type of clothing*
Grade D	Hair and where relevant the beard should be covered. A general protective suit and appropriate shoes or overshoes should be worn.
Grade C	Hair and where relevant the beard and moustache should be covered. A single or two piece trouser suit, gathered at the wrist and with high neck and appropriate shoes or overshoes should be worn. They should shed virtually no fibres or particulate matter.
Grade A/B	Headgear should totally enclose the hair and where relevant the beard and moustache; it should be tucked into the neck of the suit; a face mask should be worn to prevent shedding of droplets. Appropriate sterilised, non-powdered rubber or plastic gloves and footwear should be worn. Trouser legs should be tucked into the footwear and garment sleeves into the gloves. The protective clothing should shed virtually no fibres or particulate matter and retain particles shed by the body.

Wristwatches and jewellery should not be worn in isolator rooms because they are dirt traps and will damage gloves, sleeves and half suits. Cosmetics may generate particulate or biological contamination and should not be worn.

If staff with mild infectious diseases, such as the common cold, are obliged to work in grade C or D aseptic areas, it is recommended that they wear face masks.

Personnel responsible for spraying items into the clean room or into the isolator should wear appropriate clothing including gloves.

'Double gloving' Where a negative pressure isolator is used for aseptic hazardous products, e.g. cytotoxics, consideration should be given to the use of inner sleeves and gloves that are resistant to the substances being handled. The cuffs of the inner gloves should fit over the inner sleeves to form an effective seal.

5.6 Isolator exhaust

Gassed isolators or isolators handling hazardous materials should preferably exhaust to atmosphere with appropriate filtration, dilution or treatment to make the discharge safe. Dedicated exhaust ducts are normally specified and safety interlocks are specified where necessary. Exhaust ducts should be at negative pressure. The location of the exhaust point of the duct should comply with Health and Safety and local environmental regulations. A high discharge velocity is normally used and the typical requirement for the discharge level of a stack is 2 m above the height of the nearest building. The stack should be designed to prevent water getting into the ducting.

If exhaust to atmosphere is impractical then double filtration should be used to ensure adequate control of aerosols and particulates. If cytotoxic or other hazardous vapours might be present, then a suitable carbon filter or other appropriate measures should be used to ensure the discharge is safe.

Other aspects of isolator ducting systems are described in Chapter 2: Design. A risk assessment should be carried out to facilitate the selection and design of an exhaust system that is safe in the event of fan failure. Ducting to atmosphere will assist in controlling the level of alcohol vapour in isolator rooms.

5.7 Grade of environment

Table 5.2 has been prepared by the NHS Quality Assurance Committee in consultation with the MHRA.

It specifies the isolator pressure regime (+ve or −ve), transfer device and siting environment for a range of aseptic preparation activities. The use of terminal HEPA filtration for the isolator room may improve the air classification from Grade D to C or even B.

5.8 Segregation

Facilities should be designed so that it is easy to segregate different products and different batches of the same product. As stated in Chapter 2: Design, section 2.3.4 'Regulatory authorities advise strongly against the use of one isolator where cross-contamination represents a risk, and would normally expect to see separate dedicated facilities. If an isolator is to be used for a sequence of products, then it should be subjected to a validated cleaning process.' The following are recommendations for specific applications:

5.8.1 Cytotoxics

A dedicated isolator should be used, preferably in a dedicated room.

5.8.2 Radiopharmaceuticals

For dispensing or preparation of radiopharmaceuticals, a dedicated isolator should be used in a dedicated room.

5.8.3 Blood labelling

For handling blood, which may be contaminated with human pathogens, a dedicated isolator should be used, preferably in a dedicated room.

5.8.4 Live vaccines

For handling live vaccine and cultures, e.g. BCG, a dedicated isolator should be used, preferably in a dedicated room.

Table 5.2 Siting of isolators (prepared by the NHS Quality Assurance Committee in consultation with the MHRA)

Pharmaceutical activity	*Aseptic manipulation*	*Sanitisation*	*Isolator pressure differential*[a]	*Transfer system*[c]	*Background environment classification*[d]
Cytotoxics	Open	Gas	NA	NA	NA
		Alcohol	NA	NA	NA
	Closed	Gas	−/+	E/F	D
		Alcohol	−/+	D	D
Radiopharmaceuticals	Open	Gas	−	E/F	D
		Alcohol	−	D	D
	Closed	Gas	−	E/F	D
		Alcohol	−	D	D
Sterile non-parenteral preparations	Open	Gas	+	E/F	D
		Alcohol	+	C1/D	D
	Closed	Gas	+	E/F	D
		Alcohol	+	C1/D	D
TPN[e]	Open	Gas	NA	NA	NA
		Alcohol	NA	NA	NA
	Closed	Gas	+	E/F	D
		Alcohol	+	C1/D	D
CIVAS including antibiotics[e]	Open	Gas	NA	NA	NA
		Alcohol	NA	NA	NA
	Closed	Gas	+	E/F	D
		Alcohol	+	C1/D	D
BCG[b]	Open	Gas	NA	NA	NA
		Alcohol	NA	NA	NA
	Closed	Gas	−	E/F	D
		Alcohol	−	C2/D	D

Key
[a] Provided the isolator complies with recommendations in Chapter 2: Design.
[b] Dedicated cabinet in a dedicated room if possible.
[c] As specified in Chapter 3: Transfer devices.
[d] As specified in EC GMP Annex 1 1997.
[e] Aseptic manufacture of injections using open procedures is only applicable to UK hospital licensed units. These are covered by individual GMP inspections on a case-by-case basis.

5.8.5 CIVAS including antibiotics

A dedicated isolator in a shared room may be specified, provided traces of antibiotics are never permitted to contaminate adjacent isolators.

Penicillins/cephalosporins Because penicillins/cephalosporins can cause sensitivity reactions in operators and patients, a dedicated isolator is recommended. Special cleaning and chemical inactivation are required after manipulation. These additional precautions are extremely important.

5.8.6 CIVAS/TPN

A dedicated isolator should be used. A validated clean-down should be carried out between different product types.

5.9 Further information

Further detailed information on the design, classification, monitoring and cleaning of cleanrooms can be found in BS EN ISO 14644.

6

Cleaning, decontamination and disinfection

An isolator controlled workspace can only achieve the segregation required if it is kept free of contaminating substances, either biological or non-biological in nature. A good understanding of the principles and practical application of decontamination is essential for isolator users.

Scope

Pharmaceutical isolators require regular decontamination to facilitate product and operator protection. Decontamination normally comprises cleaning and sanitisation. These are different processes which may be combined.

This chapter outlines some of the methods and practices available, including those currently in widespread use as well as those of limited use or which are under development. To assist selection an overview is given on general methods, together with more detail on methods of choice, although this is not intended to be restrictive.

When selecting a decontamination process, the chosen process must meet the needs of and be compatible with the application. In all cases account should be taken of the four primary requirements, namely: safety, efficiency, antimicrobial activity and chemical compatibility.

For definition of terms please see Chapter 12.

6.1 Cleaning

6.1.1 General considerations

The aim of cleaning is to reduce the contamination level of a surface to render it free from dust, soil, organic or inorganic matter to a visibly clean state. Cleaning to ensure that the bioburden on surfaces is as low as possible will give a greater assurance that the disinfection or gassing process will be effective.

If performed incorrectly, cleaning activities can actually increase the levels of contamination. Where possible, cleaning should be carried out by the operators who use the isolator as they will understand the importance of cleaning, what needs cleaning and the most appropriate order of cleaning. Alternatively, cleaning may be carried out by specifically trained staff. Cleaning processes should follow validated SOPs. The effectiveness of cleaning processes should be validated, documented and regularly monitored.

Cleaning applies to the controlled workspace of the isolator, the transfer devices associated with the isolator and the background environment, including the external surfaces of the isolator. Cleaning also applies to materials entering the isolator.

Isolators are used for a large range of manufacturing, preparation and dispensing processes. For aseptic work, cleaning is always either combined with disinfection or followed by disinfection or gassing.

Opening the isolator for cleaning may give easier access for a thorough clean but must be followed by disinfection or gassing in the closed state. Isolators used for hazardous materials such as cytotoxic drug dispensing or radiopharmaceuticals, may present a risk to operators if opened without taking suitable precautions.

Considerations when designing a cleaning process are:

- level and nature of contamination of materials as they enter the isolator;
- level and nature of contamination generated by unpacking and processing;
- level of cleaning to be achieved and maintained;
- whether disinfection or sanitisation is to be included as part of the cleaning process or carried out subsequently;
- identification of all surfaces which require cleaning. These are the surfaces that may be exposed during operations including seal and gasket surfaces;
- level of acceptable residues;
- effect of cleaning on subsequent work;
- selection of cleaning agent;
- method of cleaning;
- frequency of cleaning;
- safety;
- development of SOPs;
- validation and monitoring. The effectiveness of the cleaning and disinfection processes used should be regularly validated, documented and monitored;
- relevant staff training.

Where disinfection is combined with cleaning, factors concerning the selection and effectiveness of the disinfection part of process, as described in section 6.2 should be incorporated into the combined process.

6.1.2 Cleaning schedule or timing

The isolator controlled workspace, including transfer devices, should be cleaned before and after each manufacturing, preparation or dispensing session and between activities that may result in cross-contamination.

Cleaning of general areas, such as the background environment, should be carried out at defined intervals. This may be best done during natural breaks, such as meal times, or at the end of the working day.

Every time materials are introduced into an isolator from the background environment for aseptic processing, they should undergo a surface decontamination process such as spraying and wiping with a sterile alcohol agent.

6.1.3 Cleaning methods

Isolators should be designed to facilitate easy decontamination. Some guidance on this is given in Chapter 2: Design.

Physical wiping is needed to assist in the removal of surface contamination. A progressing, overlapping, parallel wiping action should be used and circular movements avoided. Cleaning should progress in the general direction of the air flow and from top to bottom. Non-shedding, low linting cloths should be used with an appropriate cleaning agent.

Where liquid agents are used, it is important that wetting is completed on all accessible surfaces.

Special cleaning tools assist the cleaning of 'out of reach' areas when the isolator is closed. Such cleaning tools themselves must be capable of being decontaminated, and any associated swabs or pads should be disposable. Procedures should be developed to prevent the risk of a tool introducing contamination or cross-contamination due to poor technique. Cleaning materials such as wipes should be considered to be contaminated waste, which may present a hazard outside the isolator, and proper disposal procedures applied. Cleaning agents can present risks to the health of operators by inhalation or contact. Procedures should be developed to minimise these risks.

Cleaning can be done manually or by an automated process (CIP). With a manual method, it is difficult to achieve repeatability and assurance of full surface coverage, especially with complex surfaces. This provides a challenge to validation. CIP processes, as sometimes used in industrial isolators, are more easily validated.

Surfaces should be dried after cleaning and before any subsequent sanitisation process. Drying the surfaces after cleaning is important to prevent corrosion.

If surface disinfection is combined with cleaning, it should be undertaken in a closed isolator to prevent re-contamination.

Operators play a key role in the success of all manual cleaning methods. SOPs are important training and control documents. Training of operators is a requirement of GMP. A training checklist is given in section 6.9 and in Appendix 2: Training.

6.1.4 Cleaning agents

Selection of cleaning agents is important and care must be taken for process and process equipment compatibility. In all cases the requirement is to remove a surface film of contamination. Whatever cleaning agent is used, it is essential that it is used at the manufacturer's recommended dilution, and according to the manufacturer's recommended procedures. Failure to follow these guidelines may render the agent ineffective.

When preparing dilutions, the quality of the water used is important. Where cleaning agents are for use inside an isolator, water for irrigation or injection (WFI) is recommended as the diluent.

Gross soil should be removed with the aid of a detergent. The types of soils that may be present should determine the choice of the detergent formulation. Detergent cleaners are usually classified into enzymatic and non-enzymatic, with the latter usually recommended for isolators. Non-enzymatics are further classified into alkaline, neutral and acid cleaners. Alkaline cleaners are often a good choice for protein removal from a surface. The choice of a detergent will also depend on the compatibility with the isolator surfaces.

If detergents are used to remove surface films, it is important to have a final rinse stage to ensure detergent residues are removed. Rinsing can be with purified water, WFI or sterile alcohol.

Sterile alcohol is frequently used as a combined cleaning and disinfecting agent and is recommended for cleaning non-gassed isolators. It should be noted that alcohol is not sporicidal. Alcohol-based cleaning agents have been known to cause isolator enclosure materials to become 'crazed' or embrittled with time. This effect is reduced or eliminated if the surfaces are dried after treatment. Alcohol has the advantage that it removes surface films containing surfactants and some other contaminants. Alcohol may be rotated with aqueous detergents. For example, a daily or pre-process 'spray and wipe' clean with 70% alcohol may be scheduled in combination with a monthly detergent clean. (A detergent clean should always be followed by a purified water rinse to remove the detergent residues.)

Surface films can harbour contamination and interfere with disinfection and gassing efficiency. Alcohols and aldehydes can fix organic material to surfaces. Therefore any spillages containing organic material, particularly proteins, should be cleaned up with a suitable detergent before disinfection or gassing. This is an example where alcohol should not be used as an initial cleaning agent.

Use of sterile impregnated wipes instead of sprays can reduce occupational exposure and wetting of surfaces. Spray alcohols represent a theoretical fire and explosion hazard if there is inadequate ventilation. Where there is a risk of build up of alcohol vapour, suitable fans should be specified. In such situations, a risk assessment should be carried out.

Since alcohols possess no sporicidal activity, where elimination of spore contamination is required during cleaning, a sterile sporicidal disinfectant (see Table 6.3) must be used.

The EC GMP requires that disinfectants and detergents are monitored for microbial contamination. Diluted agents should be stored in previously cleaned containers for defined periods unless sterilised. Agents used to clean and disinfect Grade A and Grade B facilities should be sterile. Agents used to clean and disinfect raw materials, components and products going in or entering Grade A and Grade B facilities should be sterile.

6.2 Disinfection

6.2.1 General

Disinfection, sometimes referred to as liquid sanitisation, is a term that usually applies to the treatment of surfaces rather than volumes to achieve removal or inactivation of microorganisms to a defined level. Since microorganisms might be present on surfaces, disinfection of surfaces is of vital importance in aseptic processes. To ensure effective disinfection, all surfaces should be adequately wetted. It is generally accepted that wiping, as in cleaning (section 6.1.3), is also desirable.

Disinfection does not replace cleaning. Contamination of surfaces with chemicals or materials of biological origin may interfere with the effectiveness of disinfection. Cleaning should therefore be undertaken prior to any disinfection process or combined with it.

The decontamination of isolators used for handling blood requires special consideration. To ensure thorough decontamination after each patient's blood has been manipulated, a suitable antiviral disinfectant agent should be used. This should prevent the possibility of cross-contamination of subsequent blood. In addition, all non-disposable items that have come into contact with the blood inside the isolator, such as centrifuge buckets, gauntlets or glove-sleeves, should also be thoroughly cleaned and disinfected.

Materials being transferred into an isolator should undergo a surface decontamination process. This can be by spraying and wiping them into a transfer device or by the use of a transfer device that permits the gassing, or alternatively the sterilisation, of its contents. The method of decontamination for the transfer of materials into the isolator should be validated, documented and monitored.

Where alcohols are used, the contact times are crucial to their effectiveness. Typical values of log reduction for IMS and IPA for different contact times are given in Tables 6.1 and 6.2.

Table 6.1 Typical values of log reduction for IMS for different contact times

Test organism	*Log reduction with 70% sterile IMS at various contact times*			
	15 s	*30 s*	*60 s*	*120 s*
Staph. aureus	1.92	2.65	3.02	4.46
E. coli	1.19	1.42	2.41	3.44

Table 6.2 Typical values of log reduction for IPA for different contact times

Test organism	*Log reduction with 70% sterile IPA at various contact times*			
	15 s	*30 s*	*60 s*	*120 s*
Staph. aureus	2.77	3.09	>5	>5
E. coli	1.27	2.1	3.53	>5

After 5 min, all showed no growth with IMS or IPA.

6.2.2 Disinfectants (liquid sanitising agents)

A summary of some of the most commonly used disinfectants, their comparative effectiveness and how they kill microorganisms is given in

Table 6.3. This is not an exhaustive list. The safety, efficacy and material compatibility may vary depending on the formulation and method of application.

6.2.3 General guidance

The effectiveness of a disinfectant solution must be established before use. All disinfectant solutions must be monitored for freedom from contamination on a regular basis.

Disinfectant agents have different modes of action and different efficacies against different microorganisms. The most effective concentration of a particular disinfectant is a characteristic of that disinfectant. Effective concentrations can be widely different between disinfectants and, indeed, between manufacturers.

There is a group of European Standards for Chemical Disinfectants and Antiseptics that give laboratory test methods for antimicrobial activity. However, there are currently no universally accepted methods for surface testing of sporicidal efficacy. Therefore the selection of disinfectants should always be supported by substantiated data covering technical validation of antimicrobial action, materials compatibility, safety, shelf life and storage. This information is usually freely available from suppliers.

Relevant chemical disinfectants and antiseptics standards are:

- BS EN 1276:1997
- BS EN 1650:1998
- BS EN 13704:2002
- BS EN 13697:2001

See Chapter 11: Standards and guidelines.

6.3 Nebulisation or fogging

Liquid disinfectants can be applied to isolator surfaces by nebulisation, fogging, atomisation or spraying.

A wide range of products may be used, including hydrogen peroxide, hydrogen peroxide/peracetic acid mixtures, formalin and chlorine dioxide.

It is not easy to validate nebulisation systems. Cycle development and validation studies are conducted in a similar manner to those for gassing systems.

Nebulisation or fogging is likely to be unsuitable for most applications as there may be non-uniform coverage of surfaces, excessive liquid

Table 6.3 Disinfectants (liquid sanitising agents)

Agent type	*Bactericidal effect*	*Fungicidal effect*	*Sporicidal effect*	*Antiviral effect*	*Antimicrobial action*	*Safety*
Alcohol[a]	Very good	Good	None	Good	Disruption of cell membrane	Flammable
Stabilised chlorine dioxide[b]	Very good	Very good	Very good	Very good	Oxidising agent Cell wall damage	Toxic
Quaternary ammonium compounds and blends	Very good Gram +ve Good Gram −ve	Good	None	Mixed	Cell wall damage	Good
Phenolics	Good Gram +ve Less good Gram −ve	Good	None	Mixed	Protein coagulation	Corrosive and toxic Strong smell
Sodium hypochlorite ('bleach')[c]	Very good	Very good	Good	Good	Oxidising agent	Very corrosive and toxic
Hydrogen peroxide[d]	Very good	Very good	Very good	Very good	Oxidising agent	Corrosive
Peracetic acid/hydrogen peroxide blends	Very good	Very good	Very good	Very good	Oxidising agent	Corrosive Strong smell

Table 6.3 *Continued*

Agent type	*Bactericidal effect*	*Fungicidal effect*	*Sporicidal effect*	*Antiviral effect*	*Antimicrobial action*	*Safety*
Amphoteric blends	Very good	Good	None	Mixed	Disruption of cell membrane	Good
Glutaraldehyde[e]	Very good	Very good	Good	Very good	Alkylation of proteins and nucleic acids	Toxic Classed as a carcinogen

[a] Aqueous alcohol (70%) is the most commonly used disinfectant. Since alcohol will preserve spores, alcohol solutions should be sterilised to inactivate any spores present before use in critical zones. Sterilisation may be achieved by filtration, by the addition of a sporicidal agent such as hydrogen peroxide or by gamma irradiation.

[b] This is effective against a wide range of microorganisms and a number of common viruses. It is one of the disinfectants that may be appropriate for decontaminating isolators when blood products are being manipulated. Care should be taken not to contaminate the diluted solution and to ensure that the shelf-life is not exceeded. Chlorine dioxide can corrode metals but is safe to handle with normal precautions.

[c] This is another disinfectant that is appropriate where blood or human tissue might be present and is active against a large range of microorganisms. It is susceptible to protein contamination, which can reduce its effectiveness dramatically. It also has a very corrosive effect on metals.

[d] Solutions of hydrogen peroxide may be used for surface disinfection, but the contact time for effective disinfection is long. Formulations of hydrogen peroxide and peracetic acid act synergistically to give a more rapid antimicrobial activity than either agent on its own. These agents, individually or in combination, are also used for gaseous decontamination.

[e] Stabilised glutaraldehyde is another surface disinfectant that is more commonly used in other areas such as endoscopy. It has a very low OEL and should only be used with a dedicated extract system. Cleaning should always be carried out prior to disinfection as glutaraldehyde will chemically cross-link with soils present.

deposition and possible undesirable residues that will require cleaning after treatment.

6.4 Sporicidal gassing processes or fumigation

There is a distinction between those isolators that are designed to be gassed with a sporicidal agent and those that are not. Gassing or fumigation of the internal controlled workspaces of an isolator enables the decontamination of otherwise inaccessible spaces. The process is carried out at or near normal room temperatures so heat sensitive materials are not likely to be adversely affected.

A gassing process does not give the same sterility assurance level (SAL) as 'sterilisation'. It is important to recognise that gassing is not a sterilisation process, but a technique to reduce the bioburden inside an isolator to a defined level. A gassed isolator, when used for aseptic processing, should provide a higher level of assurance against accidental contamination of the product than a conventional cleanroom.

The selected sporicidal gassing process and its efficacy must be fully understood. Gassing processes generally consist of a number of steps forming what is termed the gassing cycle. The gassing cycle should be carefully designed, refined, documented and validated to confirm the kill level of specified, spore forming or other target microorganisms. The target microorganism should be selected taking into account the requirements of the processes to be conducted inside the isolator and any possible external or internal contamination. This is covered in greater detail in section 6.6: Biological indicators. The sporicidal gassing process should not be used routinely until it has been fully validated. Gassing cycle development and validation need a high level of expertise.

Automation of the gassing process has the advantage of repeatability resulting from the accurate control of each step. Automation makes validation easier, provides higher levels of assurance, requires less supervision and facilitates more precise planning of work schedules.

There are a number of different gassing agents. These are set out in Table 6.4 and described in detail in the following sections.

6.4.1 Formaldehyde

Formaldehyde fumigation has been used for many years. There is some documentation in the literature concerning the efficacy of the process, especially for rooms and other large volumes, but very little dealing with isolator applications. Also there are now serious concerns about its safe

Table 6.4 Sporicidal gassing or fumigation agents

Agent type	*Bactericidal effect*	*Fungicidal effect*	*Sporicidal effect*	*Virucidal effect*	*Antimicrobial action*	*Safety*
Formaldehyde	Very good	Very good	Good	Very good	Alkylation of proteins and nucleic acids	Toxic Classified as a carcinogen
Chlorine dioxide	Very good	Very good	Very good	Very good	Oxidising agent	Corrosive and toxic
Peracetic acid	Very good	Very good	Very good	Very good	Oxidising agent	Corrosive Explosion risk
Hydrogen peroxide	Very good	Very good	Very good	Very good	Oxidising agent	Refer to manufacturer

use since it is toxic and may also be carcinogenic. It can also leave harmful residues.

Humidity is vital to ensure the effectiveness of formaldehyde fumigation and should be controlled to at least 70% RH. Formaldehyde, like other aldehydes, has a chemical cross-linking mode of action and will be affected by the presence of organic soiling. Therefore surfaces should be adequately cleaned prior to fumigation.

The gas may be generated either by heating paraformaldehyde pellets ('prills'), or by heating the 37% solution of formaldehyde commonly known as 'formalin'. As an aqueous solution, formalin has the advantage of increasing the humidity. It may be good practice to boil an equal volume of water in the same chamber to increase humidity. Formaldehyde gas is poor at passive diffusion, which results in long exposure times, or the need for fan assistance to distribute the gas throughout the isolator space. HEPA filtration may have an impact on gassing cycles due to an additional biological load and gas absorption.

Removal of residual gas at the end of the cycle may be achieved by venting to atmosphere, neutralising with ammonia through suitable sterilising filters, or adsorption onto activated carbon. Whichever method is chosen, the isolator should remain closed at the end of the cycle so that contamination is not re-introduced.

Formaldehyde has been used in conjunction with steam as a sterilisation process. This is unlikely to be suitable for isolators.

6.4.2 Hydrogen peroxide vapour

Hydrogen peroxide gas has a rapid antimicrobial, including sporicidal, action and is widely used in sporicidal gassing processes for isolators. The vapour is more effective at lower concentrations than liquid hydrogen peroxide. Hydrogen peroxide gas breaks down to water vapour and oxygen and so has little adverse effect on the environment.

Gassing cycles may include an optional leak-test phase, followed by dehumidification, conditioning, gassing and aeration phases. Different suppliers may have different definitions for these phases. Dehumidification is carried out to a level of typically 40% RH or less. The gassing phase maintains the hydrogen peroxide vapour concentration or the condensation level at a defined value for a specified time. Finally, aeration reduces the hydrogen peroxide concentration to a safe level.

Hydrogen peroxide vapour is poor at passive diffusion. Gas distribution may be assisted by stirrer fans or distribution nozzles. Effective gas distribution is critical to cycle efficiency and reproducibility. All phases of hydrogen peroxide gassing cycles can be parametrically controlled, monitored and alarmed, with data recording to provide a high level of assurance of bioburden reduction, subject to prior cleaning.

Hydrogen peroxide vapour can be introduced into an isolator at a concentration up to its saturation level or dew point. If the concentration of vapour is kept below this point it is considered a 'dry' process. Alternatively, the concentration can be raised to the dew point, at which condensation will form. Biodecontamination systems that control the concentration of hydrogen peroxide vapour in this way are available commercially. The vapour is generated by flash evaporation of a metered flow of hydrogen peroxide solution into a metered and heated air stream. For each application the saturation level or dew point will be dependent on the isolator design, temperature, humidity and volume. For example, it is possible that where filters and filter plenums are not insulated, the hydrogen peroxide vapour may quickly cool and condense before reaching the main chamber. This issue should be addressed during the design of the isolator and during cycle development.

Gassing systems divide into two groups. In the open-loop group, gas is introduced to the isolator from the generator and continuously exhausted to atmosphere. In the closed-loop group, gas exhausted from the isolator is returned to the generator and there is no environmental emission. The closed-loop group further divides into 'single' or 'dual' loop versions. Single loop systems continually remove and introduce fresh vapour into the isolator to maintain a specific concentration. Dual loop systems recirculate continuously and top up the hydrogen peroxide concentration during the gassing cycle. Irrespective of the system used, the cycle must maintain the vapour level at the specified concentration in order to achieve repeatability in the level of sporicidal activity.

For all sporicidal gassing methods, only surfaces exposed to the gas or condensate are decontaminated. Ideally, all materials or components in the load should be suspended from suitable hangers and surrounded by free air space. Alternatively point contact support racks may be used. Although not a preferred practice, it may be necessary to ensure effective exposure of surfaces to gas by turning or displacing the load during the gassing cycle. This process would need to be carefully controlled and validated.

Completion of the aeration phase should be verified by measuring the hydrogen peroxide concentration. Desorption of gas absorbed on to isolator filters and condensation of gas in plenums and ducts will increase aeration times. By contrast, a high air change rate within the isolator during aeration will speed up the phase and is a consideration if a short cycle time is required. Care must be exercised to ensure that the techniques used for detecting or measuring the low concentrations of hydrogen peroxide at the end of the aeration phase are suitable. A number of commonly available detection systems have limited ranges of temperature and humidity within which they will operate.

The first gassing cycle of any isolator that is new or recently tested with DOP should be considered invalid as DOP affects the gas concentration. After this first cycle, stable and repeatable conditions are more likely and validation can start.

A recent development has been is a rapid gassing chamber. This is a transfer device in which gassing cycle times can be as low as 15 to 20 min.

6.4.3 Peracetic acid vapour

Peracetic acid gassing systems are effective but are in limited use. As with hydrogen peroxide vapour systems, liquid peracetic acid is usually heated to produce a vapour. The humidity in the space to be treated needs to be maintained and controlled at a high level to ensure an effective microbial kill. Peracetic acid can be used on its own, or as a mixture with hydrogen peroxide. Mixtures are claimed to be synergistic. The strong oxidising property of peracetic acid causes corrosion of many metals. As with hydrogen peroxide and formaldehyde, peracetic acid does not diffuse readily and its action is enhanced by using stirrer fans or other means of distributing the gas through the air space.

The peracetic process is generally a single pass process and the isolator should be vented to atmosphere. Venting may be direct or via some form of in-line treatment to reduce the gas discharge concentration to safe levels. A safety assessment should be undertaken. Peracetic acid is toxic but decomposes to acetic acid and water, both degradation products having a low toxicity.

6.4.4 Chlorine dioxide gas

Chlorine dioxide gassing systems are effective and in limited use. They have the usual cycle phases and high humidity is required. It is important that chlorine dioxide is chemically reduced or inactivated before release to the environment. The gas decomposes in the presence of light.

Chlorine dioxide gas is a broad spectrum antimicrobial agent. Sporicidal activity requires a minimum concentration of 500 ppm. Chlorine dioxide is a respiratory and mucous membrane irritant with a strong odour.

6.4.5 Ozone gas

Ozone gas generators are simple devices, and decontamination tests have been completed on pharmaceutical isolators. At the concentration required to achieve biological decontamination, ozone degrades isolator seals and gloves. It is therefore unsuitable for general use in isolator decontamination. It is also highly toxic and requires humidity control. For these reasons it is not a method of choice.

6.4.6 Ultraviolet and white light

Ultraviolet light is widely used in water treatment and may find some application in isolators. The manufacturers of UV lights claim a validated kill rate for a range of microorganisms. The effectiveness varies according to the susceptibility of the organism type, the intensity of the light source, the distance from the light source and the exposure time. The disadvantage of UV light is that kill is only achieved on surfaces that are directly exposed to the light.

A currently available commercial application is a specialised type F transfer device where UV is used to sanitise the 'ring of risk'. In this case, the port is adapted to carry an annular UV source within the front face, and the container seal and the port seal are both exposed to intense UV light immediately prior to the docking process. The exposure time slows down the transfer process appreciably.

A recent development is the use of white light. Brief pulses, less than 1 ms, of extremely intense white light have been found to inactivate a wide spectrum of microorganisms. The advantages of such a system include the absence of chemicals and of ionising radiation. Like UV light, it has been applied to the 'ring of concern' on type F transfer devices. The speed of the process does not delay the transfer, but the cost of the port is high.

6.5 Sterilisation

Sterilisation is a validated process used to render a product free from viable microorganisms.

Sterilisation can be achieved by physical and/or chemical means. Sterilisation processes include radiation, dry heat, steam, and chemical agents such as ethylene dioxide.

In a sterilisation process, the nature of microbial inactivation is described by an exponential function. Therefore the presence of a viable microorganism on any individual item can be expressed in terms of probability. While this probability, which is normally expressed as a sterility assurance level (SAL), can be reduced to a very low number, it can never be reduced to zero. SAL is the probability of a single viable microorganism occurring on or in a product after sterilisation and is normally expressed as 10^{-n}.

It is currently not necessary to ensure that isolators are sterilised, which would require validation to demonstrate and ensure a low enough SAL in every part of the isolator.

6.6 Biological indicators (BIs)

Biological indicators are dried microorganisms (generally spores), produced from standard suspensions, deposited on a carrier or coupon which is usually stainless steel and contained in a primary pack. They present a standard challenge to validate fumigation cycles. Standardisation and quality of biological indicators are crucial to cycle validation. Some biological indicator manufacturers provide baseline D values derived from testing in a reference isolator system which gives assurance of manufacturing control. The D value is the decimal reduction time. This is the time required to reduce the viable count by 1 log order, in other words to 1/10 of its previous value. Reference D-values are only applicable for the process variables used to determine them and may vary in other situations. Biological indicators should have a certificate of analysis from the supplier which must include D values or some other indicator of resistance characteristics.

Naked stainless steel coupons inoculated with microorganisms are used to simulate the hard surfaces of an empty isolator to be gassed. The coupons are best hung by wire hooks to ensure full surface exposure.

BIs, in a primary pack, can give an additional penetration challenge to the gassing process where isolators have product loads which include bagged materials or devices together with point contact support that may rely on some degree of diffusion to achieve biological inactivation.

BIs on a substrate of other materials should be used if such materials form critical contact surfaces which could present risk of product contamination. The effectiveness of a gassing process may vary depending on the contact surface. Porous materials may require specific cycle development to ensure penetration.

Bacillus stearothermophilus has been shown to be one of the most difficult microorganisms to kill with hydrogen peroxide vapour. For this reason, the pharmaceutical industry uses *B. stearothermophilus* as the BI in most filling line applications. Cycles developed using this microorganism have generally been very satisfactory.

At the time of writing, the PDA (Parenteral Drug Association) has issued a draft monograph entitled 'Recommendations for the Production, Control and Use of Biological Indicators for Sporicidal Gassing of Surfaces Within Separative Enclosures.' The final document will provide useful information about the use of BIs.

6.7 Chemical contamination

Chemical contamination may arise from accidents with or mishandling of hazardous or non-hazardous chemicals. Hazardous chemicals, such as cytotoxic drugs or active ingredients, will need special decontamination procedures. Cleaning and disinfection do not necessarily address chemical contamination.

Chemical decontamination may be achieved by decomposition, neutralisation or physical removal. To deal with chemical contamination, it is necessary to have an understanding of the chemistry. Many manufacturers of chemical products will have detailed procedures for handling spillages and for decontamination generally.

Decontamination materials such as wipes may become contaminated and present a hazard outside the isolator. Waste handling procedures must be in place with safe removal systems to contain all waste chemical hazards. These must be transported for treatment at a specialist site or equivalent in-house facility.

In addition, chemically contaminated HEPA filters from the isolator may require decontamination before handling and safe disposal.

Any decontamination process must be compatible with the pharmaceutical process and process equipment.

Isolators processing hazardous chemicals may be fitted with active carbon filters to reduce emissions to the surrounding environment. Consideration should be given to the possible degradation of carbon filters by other vapours such as those used during routine decontamination.

There are many different types of carbon filter, each meant to adsorb a particular chemical or group of chemicals. The efficiency of the filter depends on the dwell time of the airstream within the filter bed. Contaminated carbon filters, just as contaminated HEPA filters, will require safe disposal. Further information on carbon filters is given in Appendix 6: Activated carbon filters.

6.8 Validation

6.8.1 Cleaning validation

For isolators, cleaning validation is best limited to an assessment of surface contamination. This can be either visual or with a white cloth test. The white cloth should show no visible recovery when wiped over a surface following a cleaning routine. Quantitative methods with tracer chemicals, such as paracetamol or riboflavin, may also be used. Acceptance criteria for cleaning performance should be set on a case by case basis as different cleaning agents have different modes of action.

Cleaning agents should have technical data to support a validation file which should include a product specification and details of safety, compatibility, shelf-life and validation data.

It may also be necessary to validate the removal of chemical residues, including product and cleaning agent, which may provide a source of contamination that will have an impact on processed product quality.

Bioburden reduction is normally validated as part of sanitisation validation.

SOPs are required to set out the validated method of cleaning including agents recommended. SOPs should form the basis for operator training. They should contain instructions that are clear, practical and safe to put into effect.

6.8.2 Disinfection validation

Validation of disinfection methods follows similar requirements to those for cleaning and includes, in addition, verification of the removal or inactivation of microorganisms to the required level.

All critical and, where possible, key parameters of the disinfection process should be controlled, monitored and documented as evidence to support assurance of bioburden reduction.

Disinfection with liquid disinfectants may be difficult to validate in complex isolator systems and is easier with isolators having good access to internal surfaces. In all cases, the contact time of the disinfectant should be determined to provide the required level of biological decontamination. Before a new liquid disinfection procedure is accepted, it should be validated with a programme of microbiological testing.

6.8.3 Gassing validation

Validation requirements for a gassing system will depend on the design of the system and the agent used. Factors that may affect the use and validation of a particular system should be identified with the manufacturer.

The validation process is used:

- to identify the gassing cycle parameters;
- to develop appropriate gassing cycles in terms of those parameters;
- to qualify the performance of the developed cycles and thus the release criteria (the criteria which must be met before any given recorded cycle may be accepted in routine production);
- to verify that the cycle parameters (from which key data may well become the release criteria) operate within set limits.

An essential element of the cycle development process is the establishment of the time required to achieve a given log reduction of the chosen biological indicator. In order to establish this time, a 'sub-lethal' study is carried out by exposing the BIs to the gas for various times, to find out at what point in time the required log reduction is achieved. A safety margin may then be added which should be based on the observed potential for process variability or on regulatory requirements. A typical safety margin may be 50%. The resulting cycle is used in production.

Considerations for the gassing cycle development include:

- initial bioburden following the cleaning stage. This depends on the type and quantity of microorganisms observed in the isolator over time;
- materials compatibility for the isolator, process, load and equipment;
- loading and pattern;
- quality of the BIs;
- parameters of gassing cycle control provided by the gas generator;
- direct measurement of gassing conditions within the isolator (e.g. gas concentration).

Cycle development is normally preceded by studies to establish 'worst-case' (i.e. most difficult to kill) locations for the BIs within the isolator and its load. These studies include:

- flow visualisation (e.g. by DOP smoke or water mist);
- thermal mapping;
- gas distribution using chemical indicators.

Once the locations are noted, the kill time is established using BIs. This is done in the configurations in which the isolator is to be gassed in practice; that would be empty for decontaminating the isolator itself and with various standard loads in place. It is important that the loading patterns be accurately documented and then reproduced for all subsequent gassing cycles.

It is useful to know the resistance characteristics and consistency of the BIs prior to cycle development. This information is normally available from the manufacturers, though many users will carry out independent enumeration to confirm the population. It is also useful to carry out resistance studies of the batch lot of BIs used as a reference for future re-qualification, where a new batch of BIs must be used. This can be done by establishing a kill curve (also known as the death kinetics curve, a plot of BI population against gassing time) for the validated cycle in a given isolator and gas generator system.

This can subsequently be compared with the kill curve for a new batch of BIs, in the same isolator and gas generator system.

Following cycle development, three consecutive gassing cycles should be performed and validated with BIs. Some validation processes use direct inoculation of test organisms onto a surface and subsequent recovery of any survivors. This can be important as the antimicrobial effect of a decontamination process can vary depending of the surface type.

Hard copy records should be kept of all critical validation observations including documented evidence of completed cycles. Cycle records should include details of process equipment, product description and batch numbers, lot numbers of decontamination agents, cycle phase transitions (timings, concentrations of gas temperatures, humidity and any other set parameters) and successful completion of events. For regular monitoring, out of specification (OOS) deviations should be reported and appropriate actions taken. As the cycle print-out is frequently the only ‘real time’ record of the cycle, it is essential to ensure that the printer is fully functional and has an adequate supply of paper.

6.9 Training

Training of all staff is vital.

The difference between cleaning, disinfection and decontamination should be explained. Detailed instruction should be given on the principles, methods, implements, agents and safety. Safety awareness is of paramount importance when handling aggressive chemicals, and staff should receive training in dealing safely with spillages of cleaning, disinfecting and decontaminating agents. The theory of rotation of agents should be understood. Dilution and shelf-life of diluted agents should be covered.

Training in the gassing of isolators should be kept separate from other training. Training in loading the isolator, operating the gassing equipment, and dealing with error messages is essential. An understanding of microbiological controls is desirable.

Assessment of capability and a demonstration of understanding should form an integral part of the training programme. The location of SOPs and safety data sheets must be known to each member of staff.

For more information on training generally see Appendix 2: Training.

6.10 Safety

The use of agents, any of which may be hazardous to a greater or lesser extent, should be formally reviewed through protocols such as:

- risk assessment;
- HAZOP (hazard and operability studies);
- HACCP (hazard assessment of critical control points);
- FMEA (failure mode and effects analysis).

Gas monitoring devices should be used in the isolator rooms where sporicidal gassing is undertaken. These should alarm if the gas concentration exceeds the OEL (operator exposure limit) and be linked to the gas generator to shut it down.

7

Physical monitoring

The assurance that an isolator is working to its design and installation qualification will be provided by recording and monitoring the physical characteristics relating to airflows, filter integrity particulate contamination and differential pressures. The ways in which this is achieved will depend upon the configuration of the isolator.

Scope

This chapter describes physical tests used to monitor whether an isolator is performing to specification. Leak testing is quite specific to isolators and is covered in detail in Chapter 8. The remaining most common physical tests are explained in this chapter and practical guidance given. Reference is made to standards and other publications in which the tests are fully documented. Physical test protocols should already have been written at the design stage, see Chapter 2: section 2.13: Design for validation. Microbiological monitoring is dealt with in Chapter 9.

7.1 Leak testing of installed HEPA filters

Installed HEPA filters are leak tested *in situ* to ensure correct supply air and exhaust air filtration. Correct supply air filtration helps provide the required air quality inside the isolator. Other factors providing the required air quality are a sufficiently low hourly leak rate (in negative pressure isolators), a sufficient air change rate and control of process-generated contamination. Correct exhaust air filtration arrests unwanted aerosols in the exhaust air to protect personnel and the external environment, and prevents back contamination of the controlled workspace.

In Chapter 2: section 2.13 it is recommended that a draft filter testing protocol be written to ensure that suitable test ports are provided for all filters. This protocol can now be adopted.

The test is usually carried out by applying a DOP (dispersed oil particulate) challenge upstream of the filter and measuring the air quality downstream with an aerosol photometer. This is a very searching test as the upstream challenge could be as high as 10^{12} particles/m^3. By comparison, the challenge of an EC GMP Grade D background is defined as having not more than 3 500 000 0.5-micron particles/m^3 (at rest) and not more than 200 cfu m^{-3} (in operation).

All new isolators should be designed with suitable access points to introduce the DOP challenge upstream of the filter and for probes for scanning or measuring downstream. Some improvisation will be required on older equipment. For example, there may already be a pitot tube sampling point in the exhaust, which can be used for DOP sampling.

The aerosol photometer must be calibrated to draw the correct amount of air against the levels of positive and negative pressure that apply in isolators. Not all photometers are capable of doing this.

The suppliers of DOP test equipment assert that the oil is non-contaminating, and provided it is used sensibly, any residue evaporates off after testing. Nevertheless, it is essential to clean workspaces after DOP testing. Isolators utilising sporicidal gassing may have special requirements for purging prior to an operational gassing cycle.

Where possible, installed filter leak tests are scan tests and are carried out with DOP in accordance with PD 6609:2000 (in the UK) until final publication of Draft ISO 14644-3. A smoke generator is used to generate an upstream challenge of DOP which is measured by means of an aerosol photometer. The measured challenge must be within the limits defined in the standards. The photometer is set so that the measured challenge shows as 100%, which must be checked again at the end of the test. A probe is then used to scan the filter being tested. According to PD 6609:2000 the pass/fail criterion is a local penetration of 0.001% and repairs of filter media are not permitted. According to draft Draft ISO 14644-3 the pass/fail criterion is 0.01%, or as agreed by the customer and supplier, and repairs of filter media are allowed within limits. Where the tighter criterion of 0.001% is specified, the photometer must be capable of reading to these limits. In addition to the filter media, filter seals or gaskets should also be scanned if possible.

It is recognised that it is not possible to scan test all filters in an isolator, namely cartridge filters and inaccessible panel filters such as exhaust filters. A volumetric test, which measures overall penetration, may be used in these cases but unidirectional flow filters **must always** be scanned. A volumetric test for filters mounted in ducts or air-handling units is described in Draft ISO 14644-3. This may be adapted for isolator filters. Where filters are inaccessible, the test is set up and carried out in much the same way as the scan test, except that the downstream readings are taken as a series, either at a single point, or at representative defined sampling points, as mentioned in Chapter 2: section 2.5.5. The pass/fail criteria are as in the previous paragraph, but as this is for fully mixed air, the test is much less sensitive than the scan test. Even so, this test is used successfully for detecting leaks.

7.2 Particle counts

Particle counts are carried out to verify that the internal environment of an isolator and the background environment in which it is sited are both to specification. BS EN ISO 14644-1 and BS EN ISO 14644-2 give the classification of air cleanliness, sampling procedure, frequency and statistical treatment. Draft ISO 14644-3 describes the test procedure itself and the test instrument.

A light-scattering discrete particle counter is used to count and size single airborne particles at the sampling position and to report size data in terms of equivalent optical diameter.

Sampling positions should be carefully chosen. In unidirectional flow a badly chosen position may miss a contaminated airstream. In turbulent flow it may miss contamination due to dilution. However, in turbulent flow isolators, a sampling point near the exhaust will record the level of contamination in the air leaving the isolator. In the case of a continuous on-line air-sampling system, the sampling will include process-generated contamination when the isolator is in operation.

Not all particle counters have air pumps that are strong enough to be unaffected by the operational pressure of the isolator being tested. They should therefore be calibrated against the levels of positive and negative pressures that apply in isolators, to demonstrate that they draw the correct sampling volume at operational pressures.

All new isolators should be designed with suitable access points for particle counting. Some improvisation will be required on older equipment.

7.3 Airflow testing

Current clean air standards, namely BS EN ISO 14644, BS 5295:1989 and PD 6609:2000 do not specify airflow velocities or tolerances. However, EC GMP specifies a range of 0.36 m s^{-1} to 0.45 m s^{-1} (guidance value) for 'laminar flow systems' but does not give a method or grid of sampling points. It is therefore recommended that the testing of airflow velocities in unidirectional flow isolators is brought generally into line with the testing of downflow velocities in Class II

Microbiological Safety Cabinets as set out in BS EN 12469:2000 by using an appropriate measurement plane.

An anemometer is used to verify that airflow velocity in a unidirectional flow system is as specified and within the tolerance specified. A pitot tube or equivalent is used to measure airflow velocities in ducts.

Anemometers may be rotating vane, hot wire or acoustic.

Rotating vane anemometers are directional, but it should be noted that some models give the same reading whichever way the vane is rotating. If the direction of flow is opposite to that intended by the design there could be an adverse effect on process or operator safety! Rotating vane anemometers are easily damaged and therefore notorious for going out of calibration.

Hot wire anemometers are not usually directional and may be less accurate. Some instruments can be used to compute mean airflows, tolerances and air pressures.

Acoustic anemometers utilise the Doppler effect of moving air to measure both velocity and direction.

Guidelines require that airflow velocities in unidirectional flow isolators are controlled within defined limits. This is not possible in turbulent flow isolators. The next section on flow visualisation suggests a quantifiable measure for turbulent flow.

7.4 Flow visualisation and recovery testing

Flow visualisation is a procedure that utilizes different techniques to enable the airflow directions in the controlled workspace to be detected. It is usually carried out using proprietary smoke pencils, smoke guns or DOP. Smoke is used to check that the air is following an appropriate path or route inside the isolator. It is also used for checking for edge effects at mouseholes, transfer device doors or during glove breaches.

Recovery testing measures the length of time it takes for the air quality inside an isolator to return to specification after an event such as an open door, or a surge of process-generated contamination.

In unidirectional flow isolators, smoke is used to verify airflows in the unoccupied state and to observe the effect of stationary equipment, materials and transfers, with the objective of ensuring that the critical process always receives air that has come directly from the HEPA filter. It is also used to verify that any eddies or turbulence disperse quickly and that there are no standing vortices.

In turbulent flow isolators, smoke is used to observe how quickly the isolator controlled workspace is purged by the air supply and to identify any zones not swept by this air (stagnant areas). If a turbulent flow isolator is filled with smoke and then switched on, the length of time it takes for the smoke to clear can be measured. It is also possible to observe any stagnant areas where the smoke tends to linger. If the air change rate of the isolator is known, then the time taken for purging of the controlled workspace, i.e. the dwell time, can be expressed as a ratio against the time for a single air change. The ratio, which can be termed the 'clearance coefficient', can be applied to a local area of stagnation, or to the whole isolator. The clearance coefficient is a measure of air mixing or turbulence. The acceptable dwell time depends on the application

The method described in the previous paragraph is simply the time taken before the air in the isolator returns to specification. It may also be used for recovery testing for all isolators. As soon as the smoke used in the test is no longer visible, a particle count should be carried out to verify that all smoke has been purged.

A recovery test should be included as part of validation, with typical loadings in applications where good air mixing is important. Air change rates and pressures may vary within performance limits so, in critical situations, recovery testing should also be considered at the extremes of these limits.

7.5 Pressure testing

Isolators should be fitted with gauges or other indicators to enable continuous monitoring of isolator working pressure and pressure differential across the main filter, which are the key pressure readings. These pressure readings are normally alarmed to alert the operator if specified limits are exceeded.

A pressure test comprises the following:

- ensuring that all pressure gauges or indicators are reading correctly by calibration or by verification that they are within their calibration period and undamaged. It is important at the same time to check all connecting tubing for leaks and kinks;
- checking that high and low pressure alarm set points are set correctly and operate correctly;
- observing that all pressures in the isolator are within specification.

A calibrated micromanometer is used for calibration of the isolator gauges or indicators and also for measuring any pressures that are not indicated by the isolator instrumentation.

All the tests should be carried out with fresh prefilters in place. As prefilter changes are normally the responsibility of the user, it should be noted if this is being done properly.

The isolator working pressure should be checked during normal operation and during transfers to verify that it remains within the specified limits, which may be different during transfers.

The operating pressures of the transfer chambers should be checked to verify that they are correct in terms of the specified pressure cascade.

The differential pressure across a HEPA filter indicates the level of blockage. Every HEPA filter in the isolator should be checked to ensure that it has not reached the level of blockage at which the manufacturer recommends a filter change.

Fan or damper settings may need to be altered during testing to reset isolator performance to specification.

7.6 Breach testing

If a catastrophic failure should occur in the barrier of an isolator being used to process hazardous materials, there is a potential risk to the operator. If such a failure should occur in the barrier of an isolator being used to manipulate sterile materials, there is a potential risk of contamination of the sterile materials.

The well-established method of checking that an isolator system has some capability of reducing these risks is to carry out what is known as a breach test.

The breach test involves removing a single glove from an isolator in its operational state and measuring the air velocity through the resulting opening or breach. This air velocity, which is termed the breach velocity, should be at least 0.7 m s^{-1} and remain at or above this level. In critical applications, the test should be repeated at the performance limits of the isolator.

More recently it has been observed that edge effects can occur around orifices, especially if the breach velocity is significantly different from 0.7 m s^{-1}. It is therefore good practice to carry out flow visualisation tests at the opening, using smoke pencils, to verify that there is no backflow.

7.7 Recommendations for physical testing

There are five stages in the life of an isolator project when tests are carried out. These are:

- Type testing: *to establish that a particular design performs to design specification*;
- Factory testing: *to establish that a particular product performs to design specification after manufacture*;
- Commissioning: *to establish that that product performs to design specification after installation*;
- Routine maintenance testing: *to ensure that that product performs to design specification throughout its working life*;
- User monitoring: *to check that the indicated and measurable parameters remain within specification. Data should be recorded and trends monitored.*

The above tests are not validation of the isolator, although the tests may form part of the validation process. See Chapter 10: Validation.

The recommendation of the NHS QA Committee is to conduct a programme of monitoring that confirms control of the environment within standards is maintained. The correct regime for a particular isolator will depend on the application and level of usage. Refer to Beaney (2000; ref. 2 in Appendix 2: Training, p. 208) for more information.

Table 7.1 sets out a typical programme for physical monitoring tests.

All test instruments must have a current certificate of calibration.

Table 7.1 Physical monitoring tests

Test	*Type test*	*Factory test (FAT)*	*Commissioning (SAT/OQ)*[a]	*Routine maintenance test*[b] *(revalidation)*[c] *by engineer*	*User monitoring*
Gloves, sleeves, half-suits	NA	Yes	Yes	Yes	Gloves/sleeves sessionally, half-suits weekly
Installed HEPA filters	Yes	Yes	Yes	6 monthly	NA
Particle counts	Yes	Yes	Yes	3 monthly	NA
Airflow testing	Yes	Yes	Yes	3 monthly	NA
Flow visualisation and recovery	Yes	Yes	Optional	NA	NA
Pressure testing	Yes	Yes	Yes	6 monthly	Monitor continuously, record weekly
Breach	Yes	Yes	Yes	NA	NA
Alarms	Yes	Yes, calibrate set points	Yes, check set points	6 monthly, check set points	Weekly, check function
Leak testing	See Chapter 8: Leak testing, Table 8.7				

Notes

1 [a] Validation, OQ, SAT and commissioning are not the same thing. For a full explanation of these terms, refer to Chapter 10: Validation.

2 [b] Guidance on time intervals. These are generally taken from Beaney (2000; ref. 2 in Appendix 2: Training, p. 208), but note the following points:

(i) BS EN ISO 14644-2 recommends a maximum time interval of 6 months to demonstrate compliance with particle concentration limits (particle counts) for classification equal or less than ISO class 5 and suggests a maximum time interval of 24 months for what are termed optional tests. These include installed filter leakage, airflow visualisation and recovery.

(ii) For purposes of comparison there is no recommended test frequency in the current EU standard for microbiological safety cabinets. However, BS 5726-Part 4 recommends that cabinets should be regularly examined and tested by an experienced service engineer. When used in containment level 3 or level 4 laboratories (which might be considered analogous to aseptic units), cabinets may be required to be examined at least every 6 months.

(iii) A programme of testing should be established for the life time of the equipment.

3 [c] In addition revalidation should be carried out after maintenance or rectification work.

8

Leak testing

Leaks in isolators may easily be mismanaged. The guidance that follows should help in the measurement and assessment of leaks and so enable the system to be controlled and remain safe to operate.

Scope

This chapter describes how to check if leaks exist and how to measure these leaks when they are found. Methods of calculating leak rates are demonstrated and guidance values are given. Other effects that come under the general heading of leakage, namely induction effects, edge effects and macro-leakage, are described in Chapter 2: Design, and the flow visualisation used for checking such leakage is described in Chapter 7: Physical monitoring.

8.1 Introduction

The prime function of an isolator is to provide a physical barrier between the critical zone inside the isolator and the environment in which the isolator is sited. The object of this barrier may be the protection of the process, or the protection of operators, or both. Whatever the function of the isolator, the leaktightness of the structure is one of the defining factors in that function and therefore demands careful consideration. Criteria for leaktightness have already been set out in Chapter 2: Design.

When examining the leaktightness of isolators, it is important to understand that the various test methods divide into two distinct groups:

- methods that *detect* the presence and the position of leaks;
- methods that *measure* or quantify the leakage rate.

Generally speaking, it is the aim of leak testing to measure the leakage rate. If the isolator is found to leak at a rate greater than the maximum acceptable leakage rate that has been previously defined, then detection methods may be employed to locate the leak and then rectify the problem. For negative pressure isolators that are used for aseptic work, it is also recommended that a 'distributed leak test' is carried out. This is to ensure that, at commissioning and at subsequent routine testing, there are no holes that can be detected with DOP (or with helium if specified).

It should be noted that all test methods require that the isolator be sealed at a suitable test pressure that reflects its application. The isolator must therefore be engineered to make such sealing possible.

Test methods that require a reversal of pressure for testing are not recommended because:

- not all sealing systems perform equally in both directions;
- the reversal of pressure may release contamination in an unwanted direction.

It is suggested that test pressure should be a minimum of (1.5 × working pressure) but may be higher depending on the design and application of the isolator. All test instruments should have a current calibration certificate. The test and the result should be reported in an appropriate format. It should be noted that changes in atmospheric pressure and isolator temperature may affect leak rate test results. This is explained more fully in sections 8.4 and 8.5.

A single test method may not be capable of detecting a problem. Interpretation of the results of all tests is necessary to verify the performance of an isolator. A good test engineer is observant during testing and will note anything that is relevant to the performance.

Before any testing is carried out, testers should familiarise themselves with the operation of the isolator, the manufacturer's rationale for testing, and the physical location of all test points. If any test points are missing, suitable points should be devised and provided.

8.2 Leak detection methods

8.2.1 Distributed leak test

Leak detection methods are used to identify the location of leaks. However, leak detection tests may be used during commissioning to identify and cure any leaks that are capable of being detected. Such a test is called a distributed leak test. Once all leaks detected have been eliminated, the isolator can be subjected to the leak measurement test that is to be used routinely. The leakage rate measured by this test, immediately following the distributed leak test, indicates the leak rate that should be achieved in subsequent tests provided there are no significant leaks. It is recognised that leak detection tests may not detect all leaks due to inaccessibility. The helium and DOP tests described below are typically used for distributed leak tests.

8.2.2 Soap solution

The isolator is sealed at a suitable positive test pressure and soap solution is applied with a brush to potential leak areas such as seals and joints. Leaks will be identified by bubbles forming in the solution. Proprietary leak detection fluids are available.

The advantages of this method are that it is cheap and relatively effective. The disadvantages are that it is messy and labour intensive.

8.2.3 Ammonia

Ammonia solution (0.88 ammonia) is poured into a dish in the isolator, which is then sealed at a suitable positive test pressure. A test cloth impregnated with bromophenol is then moved slowly along the seams and seals of the isolator. Bromophenol is highly sensitive to ammonia and the impregnated cloth changes colour from yellow to blue on exposure.

The advantages of this method are that it is inexpensive and sensitive. The disadvantages are that it is labour-intensive and requires a skilled technician. There is a health and safety risk to be considered when handling ammonia as it has an STEL (short term exposure limit) of 35 ppm (25 mg m^{-3}).

8.2.4 Helium

'Balloon grade' helium, filtered as necessary, is introduced from a cylinder into the isolator to bring it up to a suitable positive test pressure. As helium is lighter than air, the glove/sleeve system should be agitated to ensure full mixing. A hand-held helium detector or 'sniffer' is used to scan the seams and seals of the isolator to locate any escaping helium. This test is a 'distributed leak test'. For the sensitivity of the helium test see Appendix 5: 'Calculations to estimate the size of a leak that can be detected using DOP'. The principles of the calculations apply equally to any challenge medium, including helium.

A useful way of ensuring that the helium has reached the lowest point of the isolator is to take a bleed at that point and use the sniffer to detect the presence of helium.

The advantages of this method are that it is inexpensive and non-contaminating. Helium penetrates HEPA filters; therefore the whole enclosed volume is subject to the challenge (unlike DOP, which is arrested by HEPA filters). The disadvantages are that it is labour-intensive and can only detect leaks in areas accessible to probes. The rapid dispersal of helium can make it difficult to pinpoint the leak. There is a health and safety risk as helium is an asphyxiating gas.

8.2.5 DOP (dispersed oil particulates)

DOP smoke is introduced into the isolator which is sealed at a suitable positive test pressure. A photometer is used to scan the seams and seals of the isolator for escaping DOP smoke. The DOP test is quantitative in respect of penetration. The criterion of acceptance may be taken as that for the isolator supply air HEPA filter. This may be 0.01% or 0.001% of the standard challenge according to the regulatory requirement. This test is a 'distributed leak test'. For the sensitivity of the DOP test see Appendix 5.

The advantages of this method are that engineers already have this equipment for *in situ* HEPA filter testing, the running cost of the test is low and it is very sensitive. The disadvantages are that it requires a skilled technician and can only detect leaks in areas accessible to probes. Unlike helium, it cannot detect leaks in plenums beyond HEPA filters. Also a film of DOP may be deposited in the isolator, which then needs to be cleaned thoroughly.

8.2.6 Sonics

Unlike other test methods described in this section, there is no reported experience of this method with isolators.

The isolator would be sealed at a suitable positive test pressure. Air emerging from a small hole emits a characteristic high-frequency sound. This would enable the hole to be located with a suitable detector. The detector would be used to scan the seams and seals of the isolator.

The advantage of the method is that it would be relatively inexpensive. The disadvantages are that it would require a high test pressure and would be fairly insensitive and labour-intensive.

8.3 Leak rate measurement methods

8.3.1 Introduction

Leak rate measurement is carried out:

- to verify that the leak rate is within the specified limit for the application;
- to observe trends which may indicate developing leakage before it becomes critical.

Draft ISO 14644-7 uses the four classes of containment and some of the test methods from ISO 10648-2: Containment enclosures – Part 2: Classification according to leak tightness and associated checking methods.

ISO 10648-2 expresses leakage as 'hourly leak rate' which is the ratio of the hourly leakage of the isolator to its volume.

Results can also be expressed as:

- percentage volume change per hour (see 8.6.2);
- standard decay time (see 8.6.3);
- volumetric leak rate (see 8.6.4) Note: This varies with the volume of the isolator;
- single hole equivalent (see 8.6.5) Note: This varies with the volume of the isolator.

Larger isolators have higher volumetric leak rates than smaller isolators with the same hourly leak rate. This should be taken into account when providing a rationale for a recommended maximum leak rate.

8.3.2 Factors which affect leak rate measurement

When considering leak rate measurement methods, it should be noted that there are three variable factors which may interfere significantly with their validity. These variables are:

8.3.2.1 Changes in atmospheric pressure

If atmospheric pressure changes during the test, there will be an apparent change of pressure inside the sealed isolator. Unless compensated for, such a change would affect the apparent leak rate of the isolator. One atmosphere (1000 mbar) is equivalent to 100 000 Pa. (The actual figure is 101 325 Pa, but 100 000 Pa is a perfectly valid approximation for these calculations.) A change of 1 mbar in atmospheric pressure is therefore equivalent to 100 Pa. Atmospheric pressure can change at rates of up to 0.5 mbar h^{-1}, or 50 Pa h^{-1}. Opening or closing of the door of an isolator room and adjustments to local ventilation systems may produce much larger transient changes.

8.3.2.2 Changes in temperature within the sealed isolator

The universal gas law determines that a change in temperature of 1°C within the sealed isolator will cause a pressure change of approximately 350 Pa. This is a significant effect, which must be taken into account.

$$\frac{P_1 V_1}{T_1} = \frac{P_2 V_2}{T_2}$$

$$P_2 = \frac{P_1 T_2}{T_1}$$

if P_1 is 100 000 Pa, T_1 is 20°C = 293 K and T_2 is 21°C = 294 K

$$\text{then } P_2 = 100\,000 \times \frac{294}{293} = 100\,341 \text{ Pa}$$

8.3.2.3 Movement of any flexible parts of the isolator during the test

If any flexible or elastic parts of the isolator, such as sleeves or half-suits, move during a test, the volume of the isolator will be effectively altered, and hence the internal pressure will be changed. A change in volume of 1% is equivalent to a change in internal pressure of 1000 Pa.

$$\frac{P_1 V_1}{T_1} = \frac{P_2 V_2}{T_2}$$

$$P_2 = \frac{P_1 V_1}{V_2}$$

if P_1 is 100 000 Pa, V_1 is 1 m^3 and V_2 is 0.99 m^3

$$\text{then } P_2 = 100\,000 \times \frac{1.00}{0.99} = 101\,010 \text{ Pa}$$

8.3.3 Oxygen method

This method is referenced in Draft ISO 14644-7 and described in detail in ISO 10648-2. It may be used as a manufacturer's base-line method for type-testing. The authors do not have sufficient experience of this test to comment on its usefulness.

A sensitive oxygen meter is placed inside the isolator under test, which is then filled with nitrogen, sealed, and maintained at the test pressure which must be negative. Any leakage will convey atmospheric oxygen to dilute the nitrogen in the isolator. The increase in oxygen concentration over time can be directly related to the actual leak rate at the test pressure. Other test methods can be calibrated against this test.

The advantages claimed for this method are that it is sensitive and accurate. The disadvantages are that it requires expensive equipment and a skilled technician. It also requires that the isolator be tested at a high negative pressure and with full mixing. This can be engineered if required, but many positive pressure isolators do not have these capabilities.

8.3.4 Pressure hold method

A sensitive pressure-measuring device, usually an electronic digital micromanometer, is connected to the isolator. A sensitive flow-measuring device and a small air pump with variable speed adjustment are fitted to the isolator. The isolator is sealed at the test pressure. The air pump is run and adjusted carefully to maintain the set test pressure. The air flow rate of the pump is then equal to the leak rate of the isolator.

The advantages of this method are that it is inexpensive and can be used on leaky isolator systems. It is a useful means of instantly observing the effects of remedial leak repair. The disadvantages are that it has not been widely evaluated for isolators and the necessary flow rate equilibrium may be difficult to achieve. Also, it is affected by changes in atmospheric pressure and isolator temperature.

8.3.5 Parjo test method

This method (see Figure 8.1), adopted in Draft ISO 14644-7, has been extensively described in nuclear industry publications and standards. A rigid sealed reference vessel, such as a 2-litre glass bottle, is placed inside the isolator. This is fitted with a glass tube, bent through 90° to provide a short horizontal section. A soap bubble is introduced into

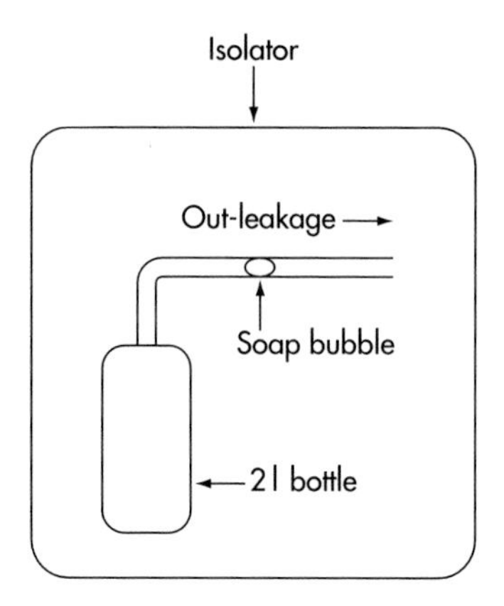

Figure 8.1 Parjo leak test.

this horizontal section. The isolator is sealed at the test pressure, either positive or negative. Any outward leakage during a positive pressure test will cause the bubble to move away from the vessel neck and vice versa. If the bore of the tube is known, the rate at which the bubble moves can be measured and used to calculate the actual leak-rate of the isolator.

The advantages of this test are that it is cheap and quick. As the test is affected by changes in atmospheric pressure and isolator temperature, it should be repeated until consistent results are obtained. The disadvantages are that it is only suitable for relatively small leak rates and requires a skilled technician. This method is not known to have been used in pharmaceutical isolators. This and other reference vessel methods may be developed for future practical application.

8.3.6 Pressure decay method

This method is referenced in Draft ISO 14644-7 and described in detail in ISO 10648-2. It has been used extensively for pharmaceutical isolators. The isolator is sealed at the test pressure differential and the decay of that pressure differential measured over time with a device such as a micromanometer. The observed rate of pressure decay converts directly to the leak rate. ISO 10649-2 specifies a test time of 1 hour, but shorter test times are considered appropriate for pharmaceutical isolators.

The advantages of the method are that it is cheap and can be automated. The disadvantages are that it is affected by changes in atmospheric pressure, isolator temperature and isolator volume.

8.4 Discussion of pressure decay leak rate measurement

The simplicity of the pressure decay test has made it the most widely used of the leak measurement methods and the following means have been developed to minimise the effects of the three variables previously described.

8.4.1 Changes in atmospheric pressure

For isolators where the maximum permissible leak rate is relatively high, the test can be carried out very quickly and the effect of changes in atmospheric pressure ignored. The test is carried out in accordance with the first method described in section 8.5.

For isolators where the maximum permissible leak rate is relatively low and the test is therefore prolonged, it is possible to compensate by measuring the atmospheric pressure at the start and finish of the test and applying a correction. This can be done in accordance with the second method described in section 8.5.

For isolators where the maximum permissible leak rate is extremely low, test pressure differentials may be increased to 1000 Pa. This has the effect of revealing small leaks more quickly. However, many isolators are designed to operate at pressures of no more that 120 Pa and are therefore not necessarily capable of withstanding test pressures of 1000 Pa.

8.4.2 Changes in internal temperature of the isolator

For isolators where the maximum permissible leak rate is relatively high, the test can be carried out very quickly and the effect of changes in internal temperature ignored. The test is carried out in accordance with the first method described in section 8.5.

For isolators where the maximum permissible leak rate is relatively low and the test is therefore prolonged, it is possible to compensate by measuring the internal temperature at the start and finish of the test and applying a correction. This can be done in accordance with the second method described in section 8.5.

8.4.3 Movement in flexible parts

For all pressure decay tests, glove sleeve systems should be supported to reduce changes in shape and volume of the isolator during the tests or, better still, capped off and leak-tested separately.

For isolators manufactured from flexible film and, to a lesser extent, rigid materials, it must be recognised that the effect of a leak may result in both a change in volume and a change in pressure differential. Therefore the actual decay in pressure differential will not be as great as if the isolator were absolutely rigid, and the actual rate of leakage may be understated. This effect should be taken into account when setting the maximum recommended leak rate.

8.5 Guidelines for leak rate measurement by pressure decay

Ideally, for all pressure decay tests the isolator should be in thermal equilibrium with its surroundings. The isolator may gain heat from such sources as lights, controls and fans. Therefore these should be switched off well ahead of a test. Heat may also be gained from sunlight and room radiators, and the effect of these must be minimised prior to and during the test. There should be no changes in pressure in the isolator room during the period of the test; for example, doors to other rooms of differing pressure must not be opened.

8.5.1 Test method for isolators where a maximum leak rate of 1.0% per hour is specified

Prepare the glove/sleeve systems by placing them in a position where they are least likely to move during the test. In positive pressure isolators this is achieved either by everting the sleeves and supporting the cuff rings or by inserting rigid or inflatable supporting devices to restrain movement. Such devices may also be used in negative pressure isolators, or the cuff rings may be supported from the work surface of the isolator.

1. Switch on the fan or blower of the isolator to raise (or lower) the internal pressure. Seal off the isolator using the valves or test lids provided for the purpose, closing last the valve or lid which 'traps' the test pressure inside the enclosure. Then switch off the fan or blower.
2. Observe the internal pressure of the isolator and note the time at which it reaches a predetermined value. This may be a standard test pressure of 250 Pa or any other suitable pressure, e.g. a pressure of 1.5 × normal operating pressure. See also section 8.7.
3. Note the time taken for the isolator pressure differential to decay by 25 Pa. Express the rate of decay of differential pressure in the required form for reporting. To pass this test the rate of pressure

decay must be <25 Pa in 1.5 min, equivalent to a leak rate of <1.0% volume loss per hour. This time is more practical for this leak rate that the 1 hour specified in ISO 10648-2.

8.5.2 Test method for isolators where a maximum leak rate of 0.25% or 0.05% per hour is specified

Prepare the glove sleeve systems in accordance with Test Method 8.5.1 above.

1. With the isolator in thermal equilibrium with its surroundings, switch on the fan or blower of the isolator to raise (or lower) the internal pressure. Close and seal the isolator using the valves external to the inlet and exhaust HEPA filters.
2. Raise the pressure of the isolator to a few pascals above the specified test pressure, check that all sleeves and gloves are stable and wait 10 min.
3. When the pressure has stabilised, start the test and note the starting pressure, the time, the atmospheric pressure (to an accuracy of 0.01 mbar) and the temperature (to an accuracy of 0.01°C) inside the isolator. During the period of the test, events which may cause a change of room pressure, such as opening doors to rooms with different pressures, should be avoided.
4. After the specified test period, note the isolator pressure, the isolator temperature and atmospheric pressure.
5. Correct the pressure decay figure for any variation in atmospheric pressure. (If atmospheric pressure has gone up, add 1 Pa to the final pressure reading for every 0.01 mbar and vice versa). Figures 8.2 and 8.3 show the effect of a change in atmospheric pressure of 1 mbar (100 Pa) on the pressure reading of a non-leaking positive

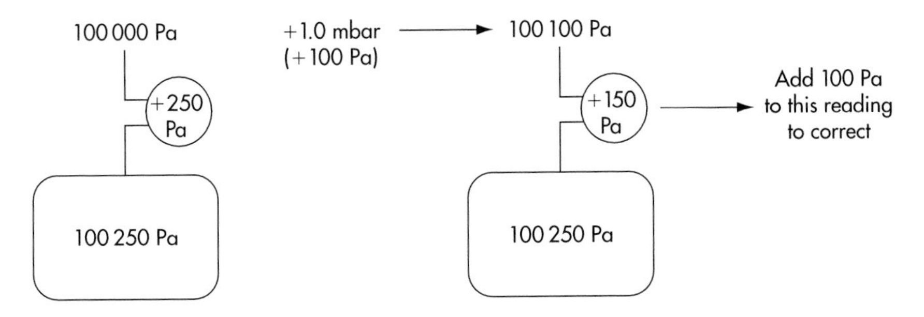

Figure 8.2 Positive pressure test: correction for atmospheric pressure rise.

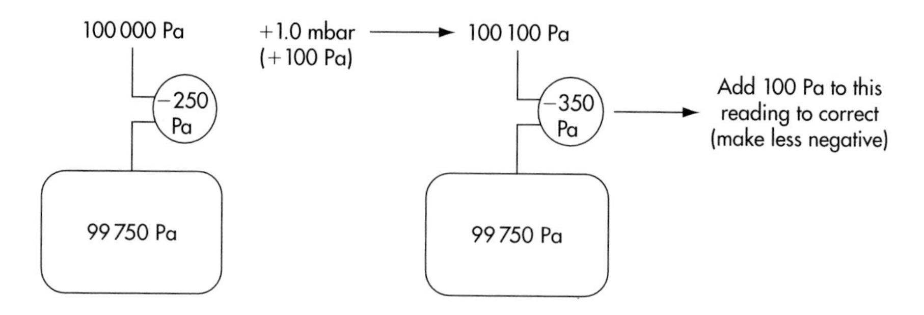

Figure 8.3 Negative pressure test: correction for atmospheric pressure rise.

isolator and a non-leaking negative isolator, respectively. The correction is shown in each case. In an actual pressure decay test, the atmospheric pressure is measured at the start and finish of the test and the correction applied in the same way.

6. Correct the pressure decay figure for any variation in temperature in the isolator. (If temperature has gone up, subtract 3.5 Pa from the final pressure reading for every 0.01°C and vice versa.) Figures 8.4 and 8.5 show the effect of a change in isolator temperature of 1°C (350 Pa) on the pressure reading of a non-leaking positive isolator and a non-leaking negative isolator, respectively. The correction is shown in each case. In an actual pressure decay test, the internal isolator temperature is measured at the start and finish of the test and the correction applied in the same way.
7. Express the pressure decay figures in the required form for reporting. To pass this test, the rate of pressure decay must be <25 Pa in 6 min, equivalent to a leak rate of <0.25% volume loss per hour, or <25 Pa in 30 min, equivalent to a leak rate of 0.05% volume loss per hour. These times are more practical for these leak rates than the 1 hour specified in ISO 10648-2.

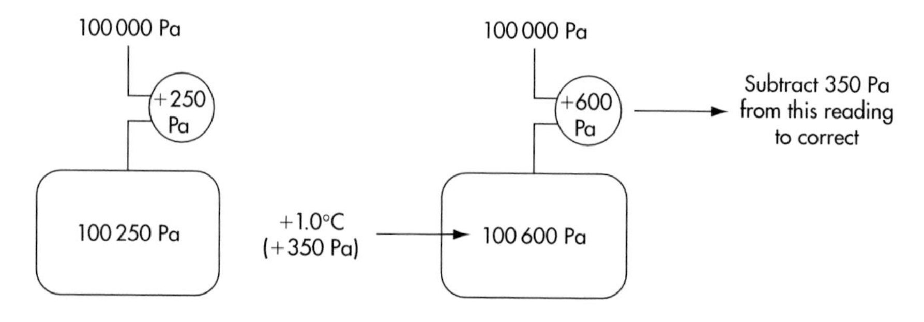

Figure 8.4 Positive pressure test: correction for internal temperature rise.

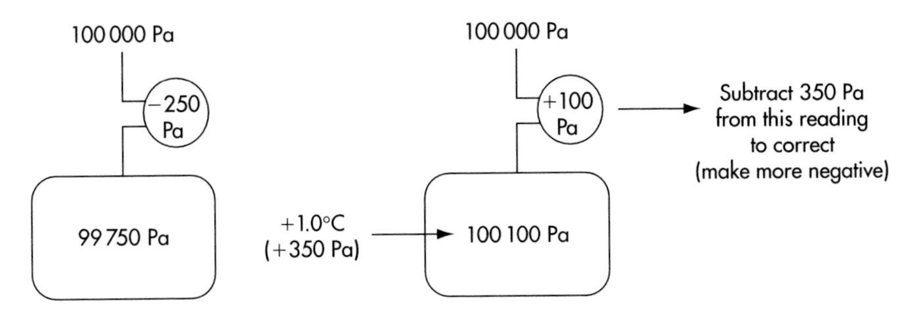

Figure 8.5 Negative pressure test: correction for internal temperature rise.

8.6 Expression of pressure decay and pressure hold leak rate results

The results from the pressure hold and pressure decay leak test measurements may be expressed in a variety of ways, but the following are suggested as standardised expressions:

8.6.1 Hourly leak rate

Expressed in h^{-1}

This is the expression used in Draft ISO 14644-7 and ISO 10648-2. It is defined as the ratio between the hourly leakage of the isolator and its volume. It is expressed in reciprocal hours (h^{-1}).

Draft ISO 14644-7 and ISO 10648-2 specify four classes of containment (see Table 8.1) with different hourly leak rates ‘measured at the normal operating pressure (usually about 250 Pa) for checking during operational use, and up to 1000 Pa for the acceptance test’. Note: Pharmaceutical isolators operate at pressure differentials upwards of 50 Pa.

Table 8.1 Hourly leak rates

Class	*Hourly leak rate* (h^{-1})	*Pressure integrity*	*Test methods*
1	$\leqq 5 \times 10^{-4}$	High	Oxygen, pressure change or Parjo
2	$<2.5 \times 10^{-3}$	Medium	Oxygen, pressure change or Parjo
3	$<1 \times 10^{-2}$	Low	Oxygen, pressure change or constant pressure
4	$<1 \times 10^{-1}$		Constant pressure

Class 4 is not considered in this book.

Hourly leak rate gives an indication of the theoretical overall effect of the leak and is an expression that is easy to use for formulae and calculations.

Hourly leak rate can be measured by means of the pressure decay test. It might seem obvious that the hourly leak rate should be the same as the hourly pressure decay rate and this can be demonstrated mathematically as follows:

The gas law states that

$$\frac{P_1 V_1}{T_1} = \frac{P_2 V_2}{T_2}$$

where P_1 is the initial absolute pressure
V_1 is the initial volume
T_1 is the initial absolute temperature

and P_2 is the final absolute pressure
V_2 is the final volume
T_2 is the final absolute temperature

If the temperature remains constant, $T_1 = T_2$,

then $P_1 V_1 = P_2 V_2$

$$P_2 = \frac{P_1 V_1}{V_2}$$

Now the hourly leak rate $n = (V_2 - V_1)/V_1\ \mathrm{h}^{-1}$, so

$$V_2 = V_1(1 + n)$$

(This is because a quantity of air nV_1 has leaked out of the isolator and the pressure inside responds (reduces) as if the volume were greater by the volume of air that has leaked out.)

$$P_2 = \frac{P_1 V_1}{V_1(1 + n)}$$

$$\therefore P_2 = \frac{P_1}{(1 + n)}$$

$$\therefore P_2 + P_2 n = P_1$$

$$\therefore P_2 n = P_1 - P_2$$

$$\therefore n = \frac{(P_1 - P_2)}{P_2}$$

which is the rate of decay in absolute pressure.

8.6.2 Percentage volume change per hour

Expressed in % of isolator volume lost per hour or % h^{-1}

This expression is the volume of air lost (or gained) from the isolator per hour, expressed as a percentage of the volume of the isolator, at the test pressure. For the three classes of isolator considered, these values are $\leqq 1.0\%\ h^{-1}$, $<0.25\%\ h^{-1}$ and $<0.05\%\ h^{-1}$.

Percentage volume loss per hour is a readily understood indication of the extent of the leak and is simply the hourly leak rate expressed as a percentage.

8.6.3 Standard decay time

Expressed in minutes

This is the time taken for the isolator to drop 25 Pa from a standard test pressure of 250 Pa and may be used to place the isolator in its appropriate class of $\leqq 1 \times 10^{-2}\ h^{-1}$, $<2.5 \times 10^{-3}\ h^{-1}$ or $<5 \times 10^{-4}\ h^{-1}$ as shown in the calculations that follow.

Note: In the following calculations it is a sufficient approximation to take the isolator pressure as being 100 000 Pa.

If n is the hourly leak rate in Pa^{-1}

$$\text{Pressure drop} = n \times 100\,000\ \text{Pa h}^{-1}$$

$$= \frac{(n \times 100\,000)}{60}\ \text{Pa min}^{-1}$$

$$\text{Decay time (min Pa}^{-1}\text{)} = \frac{60}{(n \times 100\,000)}$$

$$\text{Standard decay time (min (25 Pa)}^{-1}\text{)} = \frac{60 \times 25}{(n \times 100\,000)}$$

See Table 8.2.

Table 8.2 Relating standard decay time to hourly leak rate

Class of isolator (Draft ISO 14644-7 & ISO 10648-2)	*Hourly leak rate (h^{-1})*	*Standard decay time (min)*
3	10^{-2}	$\frac{60 \times 25}{10^{-2} \times 100\,000} = 1.5$
2	2.5×10^{-3}	$\frac{60 \times 25}{2.5 \times 10^{-3} \times 100\,000} = 6.0$
1	5×10^{-4}	$\frac{60 \times 25}{2.5 \times 10^{-4} \times 100\,000} = 30$

8.6.4 Volumetric leak rate

Expressed in $m^3 s^{-1}$

This is the volume of air lost (or gained) from the isolator per second expressed as volume per second at the test pressure.

Isolators of different sizes but with the same hourly leak rate will have different rates of volume loss per second.

For an isolator whose volume is V m^3

$$\text{Volume loss per second} = \frac{V \times n}{3600} \text{ m}^3 \text{ s}^{-1}$$

See Table 8.3.

Table 8.3 Relating volumetric leak rate to hourly leak rate for an isolator of 1 m^3

Class of isolator (Draft ISO 14644-7 & ISO 10648-2)	*Hourly leak rate (h^{-1})*	*Volumetric leak rate ($m^3 s^{-1}$)*
3	10^{-2}	$1 \times \frac{10^{-2}}{3600} = 2.8 \times 10^{-6}$
2	2.5×10^{-3}	$1 \times \frac{2.5 \times 10^{-3}}{3600} = 0.70 \times 10^{-6}$
1	5×10^{-4}	$1 \times \frac{5 \times 10^{-4}}{3600} = 0.14 \times 10^{-6}$

8.6.5 Single hole equivalent (SHE)

Expressed in micrometres (microns)

This is the diameter of the theoretical single circular hole, in a wall whose thickness is less than the diameter of the hole, which would account for all of the observed leakage. Further assumptions are made in the following calculations of SHE with regard to aerodynamics, viscosity and temperature. In practice the leakage is likely to be through a number of tortuous pathways. SHE is therefore very much a comparative expression rather than a real value.

Single hole equivalent was originally conceived as giving an indication of the size of the hole in relation to the size of particle that might pass through the hole (in either direction). For the reasons given in the previous paragraph, it is now recognised as being no more than an index of isolator leaktightness.

Isolators of different sizes but with the same hourly leak rate will have different values of SHE.

The SHE is calculated from the volume flow rate through the leak and the velocity of the leak. The volume flow rate of the leak is the hourly leak rate (as measured by the pressure decay test) × the volume of the isolator.

The velocity of the leak is derived from the pressure differential across the leak. The following calculation is an approximation as it is for the 'effective' rather than the actual hole diameter and it ignores the effect of temperature and pressure:

$$\text{Velocity of leak} = \sqrt{\frac{2\Delta P}{\rho}}$$

$$\text{Average } \Delta P = \frac{(250 + 225)}{2} = 237.5 \text{ Pa}$$

$$\rho = 1.225 \text{ kg m}^{-3}$$

The value for air density, ρ, varies with altitude, moisture content and temperature.

$$\text{Velocity of leak} = \sqrt{\frac{2 \times 237.5}{1.225}} = 16.07 \text{ m s}^{-1}$$

$$\text{Area of hole, } a = \frac{(\text{volume flow rate})}{\text{velocity}} \text{ m}^2$$

See Table 8.4.

Table 8.4 Relating single hole equivalent (SHE) to hourly leak rate for an isolator of 1 m^3

Class of isolator (ISO 14644-7 & ISO 10648-2)	*Hourly leak rate (h^{-1})*	*Leak rate ($m^3\ s^{-1}$)*	*Area (m^2)*	*SHE (diameter)*	
				(m)	*(μm)*
3	10^{-2}	2.8×10^{-6}	0.17424×10^{-6}	0.464×10^{-3}	464
2	2.5×10^{-3}	0.7×10^{-6}	0.04356×10^{-6}	0.232×10^{-3}	232
1	5×10^{-4}	0.14×10^{-6}	0.0087×10^{-6}	0.103×10^{-3}	103

8.7 Summary of expressions

Tables 8.5 and 8.6 consolidate the different expressions for leaks derived in section 8.6.

The values in Table 8.6 have been calculated for an isolator of 1 m^3 volume and should be re-calculated for different volumes of isolator.

Table 8.5 Relating percentage volume change per hour and standard decay time to hourly leak rate

Class of isolator	*Hourly leak rate (h^{-1})*	*Percentage volume change per hour (% h^{-1})*	*Standard decay time (for 25 Pa drop) (min)*
3	$\leqq 1 \times 10^{-2}$	$\leqq 1.0$	>1.5
2	$<2.5 \times 10^{-3}$	<0.25	>6
1	$<5 \times 10^{-4}$	<0.05	>30

Table 8.6 Relating volumetric leak rate and single hole equivalent to hourly leak rate calculated for an isolator of 1 m^3 volume

Class of isolator	*Hourly leak rate (h^{-1})*	*Volumetric leak rate ($m^3\ s^{-1}$)*	*Single hole equivalent (SHE) (μm)*
3	$\leqq 1 \times 10^{-2}$	2.8×10^{-6}	464
2	$<2.5 \times 10^{-3}$	0.70×10^{-6}	232
1	$<5 \times 10^{-4}$	0.14×10^{-6}	103

Note: These values should be re-calculated for different volumes of isolator.

8.8 Determination of acceptable leak rates

The acceptable leak rate for any individual isolator system should be determined, taking into account its configuration, volume, application, risks to process and hazards to operators. Isolator manufacturers should give a rationale for the acceptable leak rate for their isolator for a specific application. The mass balance method in Draft ISO 14644-7 Appendix E Section E.3.2 can be used for estimating the effect of a leak. The method estimates the concentration of airborne contamination downstream of a leak when the upstream concentration, volume flow rate of the leak and air change rate of the space downstream of the leak are all known (see Figure 8.6). Although the method assumes perfect mixing, which is not realistic, it gives a useful indication of the effect of a leak into a negative pressure isolator or into a room from a positive pressure isolator. A leak test should be specified from Draft ISO 14644-7, ISO 10648-2, or these guidelines. Alternatively, manufacturers may propose any other method that is appropriate for their isolator. Such a method must be supported by a scientific rationale.

8.9 Tests for gloves, sleeves and half-suits

8.9.1 Gloves

The gloves used in an isolator generally constitute the area of greatest hazard associated with leakage. They tend to be relatively weak, they are flexed heavily and continually in use, and they are likely to be in

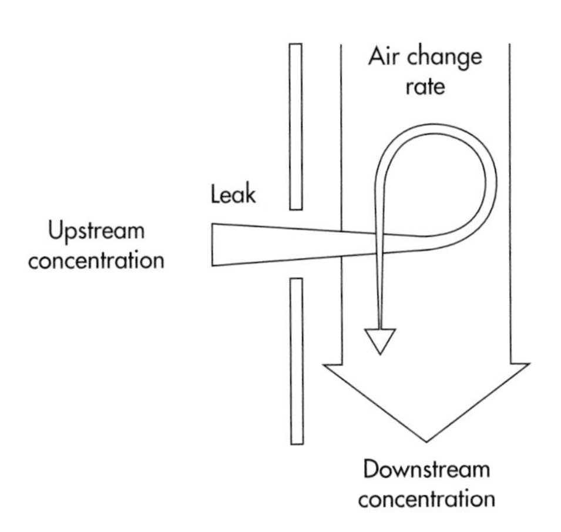

$$\text{Downstream concentration} = \frac{\text{Upstream concentration} \times \text{Volume flow rate of leak}}{\text{Air change rate} \times \text{Volume of space}}$$

Figure 8.6 Mass balance method for estimation of acceptable leak rate into or out of isolators.

close proximity to the work in hand. For these reasons, gloves need special attention in order to reduce the risks associated with leakage. The following precautions should be considered:

1. In applications where new gloves are supplied for non-sterile isolators, they may be tested for leaks before being fitted to the isolator. A rig may be devised whereby gloves are inflated and immersed in water to visualise leaks. High test pressures of at least 250 Pa must be used since gloves have a self-sealing tendency.
2. Gloves already in use should be inspected visually for tears, cuts or nicks, on a regular basis, ideally before and after every work session.
3. Gloves already in use may be tested by proprietary in-use testers. One example of such a tester requires the glove to be drawn out into a small chamber where it is surrounded with a nitrogen atmosphere at negative pressure. The test detects oxygen entering the chamber from any glove leak.
4. Gloves may be tested as part of a complete glove/sleeve assembly (see section 8.9.2 Sleeves).
5. Gauntlets (i.e. single piece gloves and sleeves) should be treated in the same way as simple gloves.

8.9.2 Sleeves

Isolator sleeves are the next most vulnerable part of isolators in terms of leakage and should be inspected and tested as follows:

1. As with gloves, regular visual inspection should be carried out. Sleeves may be constructed to indicate leakage by the clear separation of two layers.
2. The sleeve and glove assembly may be leak tested *in situ* with a proprietary test device. Such a device usually comprises a disc which seals the sleeve at the shoulder ring. For positive isolators the disc is placed into the shoulder ring from inside the isolator. For negative pressure isolators the disc is placed into the shoulder ring from outside the isolator. The pressure decay of the assembly can either be observed as the assembly droops or be measured with a micromanometer. Where test devices incorporate pumps to increase the inflation pressure, consideration should be given to the potential contamination of the isolator in the event of a leak being present. The sleeves should be tested to at least the same standard as the body of the isolator.

8.9.3 Half-suits

Half-suits are similar to sleeves in terms of leak hazard and should be inspected and tested as follows:

1. As with sleeves, regular visual inspection is recommended. Sleeves may be constructed to indicate leakage by the clear separation of two layers.
2. Ideally, new suits should be leak-tested before fitting and a certificate provided by the supplier.
3. Proprietary in-service leak test devices may become available. Currently some of the test methods used for isolators may be applied to half-suits.

Where sleeves and suits are found to have failed, repairs may be made as an interim measure. Some manufacturers provide a repair kit. Any repairs made to these access devices should be tested for leakage. Consideration should be given to re-sanitisation of the isolator following such repairs.

8.10 Leak testing schedule

A typical leak testing schedule is given in Table 8.7. For standard isolators, leak testing is also carried out as part of the type test to confirm integrity of design.

Table 8.7 Typical leak testing schedule

Test	*Factory test (FAT)*	*Commissioning (SAT/OQ)*	*Routine maintenance tests (revalidation) by engineer*	*User monitoring*
Distributed leak test, e.g. DOP or helium	Yes	Yes	N/A	N/A
Leak rate measurement, e.g. pressure decay	Yes	Yes	6 monthly	Positive isolator monthly Negative isolator weekly
Leak detection	Used as distributed leak test before or during FAT and SAT		6 monthly[a]	
Gloves/sleeves, e.g. pressure decay	Yes	Yes	6 monthly	Before each working session
Half-suits, e.g. pressure decay	Yes	Yes	6 monthly	As for isolator

Note: Leak testing should also be performed after non-routine maintenance where integrity may have been compromised.
[a] Used if the measured leak rate is in excess of specification.

9

Microbiological monitoring

Monitoring of isolators for compliance to an operational requirement is sensibly divided into physical tests and microbiological tests. This section addresses the microbiological monitoring required for viable micro-organisms.

Scope

This section provides guidance for the microbiological monitoring of isolators. The frequency of testing and the methods recommended represent good practice. However, they are not prescriptive and may require modification for individual situations to ensure that adequate monitoring levels are in place. A suitably qualified and experienced person should be appointed to review the monitoring programme. This person is likely to be the person responsible for microbiological monitoring on the site.

9.1 General

9.1.1 Media

It is important that the most appropriate media are used for the culture of samples. Two commonly used semi-solid growth media are identified in Table 9.1 and their applications shown.

It should be noted that as well as fungi, yeast and moulds, bacteria will also grow on SAB and can be mistaken for non-filamentous moulds. Also, if large numbers of bacteria are expected, then a fungal count

Table 9.1 Applications of two commonly used semi-solid growth media

Type of media	*Application*
Tryptone soya agar (TSA)	Total viable counts
Sabouraud dextrose agar (SAB)	Fungi

should be carried out using SAB with a suitable bactericide such as chloramphenicol. The bactericide will suppress bacterial growth that could mask fungal growth with which it competes for space and nutrients.

9.1.2 Fertility challenge, sterility and records

A suitable system should be in place which can be used to ensure the ability of the culture media to support the growth of the likely contaminants and also that the media is sterile on receipt. This should be confirmed by a certificate of sterility and fertility.

Media used in critical zones, such as isolator controlled workspace and transfer devices, should be sterilised to prevent the introduction of contamination into these zones. Sterilisation of the media by irradiation is a suitable method. The process of sanitisation must not impair the ability of the media to support growth. In particular, it is imperative that monitoring plates used within gassed isolators are protected by gas-resistant packaging until required. The use of sterilised media also reduces the incidence of false positive growths being detected.

Any items used for microbiological monitoring should be surface decontaminated or sanitised before being passed into the isolator or transfer device. Sterilised media is usually available as a double wrapped package to enable aseptic transfer into the isolator controlled workspace before unwrapping.

A visual check for contamination should also be carried out on all microbiological media prior to passing it into the controlled workspace.

All media used should be obtained from a reliable source and traceable to that source with a batch number, expiry date and certificate of fertility and sterility. Records should be maintained.

Subsequent purchase of more batches or brands of material should be from a source which complies with a similar standard. Sensible purchasing should reduce unnecessary variance in sampling efficiency. If media is to be produced in-house, all necessary checks should be carried out before it is put into use.

9.1.3 Sampling

When a set of samples is taken according to established SOPs it is recommended that the following data be recorded:

- location or location code;
- date and time;

- exposure time in the case of air sampling;
- activities, products produced or procedures taking place in the unit;
- identities of operators carrying out the main activities in the isolator;
- identities of other staff used for ancillary duties such as surface decontamination;
- identities of sampling staff.

A record of any previous incidents or problems should be accessible.

A plan or schematic drawing of the isolator should be obtained or prepared. The person responsible should mark on the drawing the location of sampling sites for each of the different types of monitoring test. A reference code should be used for each sampling point. These location codes should be understood by monitoring staff and they should form part of the relevant SOP. The codes should indicate where the following tests are to be done:

- settle plates;
- active air sampling;
- surface sampling;
- finger dabs.

It is important to adhere to the sampling plan so that the results can be analysed in a meaningful manner. If there is poor adherence to the sampling plan, it will not be possible to derive any trends for the data.

As sampling in fixed positions on a regular basis may not reveal microbial contamination in un-sampled zones, the SOP should also include periodic random pattern sampling in different locations to ensure such contamination does not remain undetected.

If it is necessary to investigate incidents of unusual contamination, additional sampling will be required to help with the investigation. This additional sampling should form part of the SOP.

9.1.4 Culture incubation and recording

All media, in whatever form, should be retrieved promptly and transferred to the incubator for culture. Recommended incubation times and temperatures for most common environmental contaminants are given in Table 9.2.

There are different recommendations for the best temperatures for incubation of naturally occurring environmental organisms. With TSA it

Table 9.2 Recommended incubation times and temperatures for most common environmental contaminants

Medium	*Incubation temperature (°C)*	*Incubation time to initial visual check (h)*	*Total recommended incubation time (days)*
Tryptone soya agar (TSA)	30–35	48	7–14
Sabouraud dextrose agar (SAB)	20–25	48	7–14

is probably better to incubate at 32°C as this may be the best temperature for the quick recovery (revival) of stressed, transient and commensal organisms, which may be present in clean rooms. (A stressed organism is one that is damaged but not killed by a sanitising agent or heat; a transient organism is not typically resident in the background environment but is introduced by equipment, packaging or people; a commensal organism lives in close association with people.) A temperature of 37°C may not culture these environmental organisms effectively as it is too hot, but is commonly used for culturing pathogenic organisms.

Exposed plates may be over-wrapped in sealed plastic bags to prevent desiccation. Alternatively, the humidity of the incubator may be controlled to prevent undue drying or desiccation of the plates as this will also inhibit recovery (revival) and growth of microorganisms. This is more important for longer incubation times as some plates may dry out if incubated for 14 days in some incubators. This is especially true of plates used to sample laminar flow environments because they will have already started to dry out in the airflow. The effects of desiccation can be lessened by using deeper poured plates.

Plates should be examined for growth at intervals during the incubation period. The SOP should include a schedule for checking plates. The checks may be daily or, for example, after 2, 3, 4, 5, 7, 10 and 14 days.

Fungal spores from a fungal growth from a sample may contaminate an incubator. Plates showing fungal growth should be removed from the incubator on discovery to prevent this. Fungi grow well at normal room temperatures so further incubation could be carried out on the bench. However, once the presence of fungi has been detected, further incubation is not necessary. Indeed, the growth of one fungal organism is likely to give rise to many more, so prolonged incubation encourages growths from spores from the growing colony itself rather than from the area sampled.

Different microbiologists may have different views on incubation temperatures, incubation times and precautions regarding fungal spores.

9.1.5 Reporting

Microorganisms may occur in haploid (singular), diploid (pairs), chain and bunch form, and any of these can give rise to a single colony. The only practical way of quantifying the microorganisms present in the sample is to count the colonies generated. The term cfu (colony forming unit) is used.

The colonies are counted, recorded as raw data and expressed as follows:

- air sample: cfu/m^3;
- settle plates (diameter 90 mm): cfu/4 hours;
- contact plates (diameter 55 mm): cfu/plate;
- glove print (5 fingers): cfu/glove.

Other expressions may be used, but consistency is necessary to facilitate meaningful comparison with other monitored zones.

Electronic storage of data on a suitable database or spreadsheet recording system will enable reporting in an efficient and logical manner.

The analysis of data to reveal trends is important. As many of the results will be numerically low or zero in value, the data analysis software should be capable of handling low values. Appropriate trending analysis by computer should reveal significant changes in contamination levels.

Recommended limits or guidance limits for contamination levels are given in the EC GMP and Beaney (2000; ref. 2 of Appendix 2, p. 208). The person responsible will be expected to review results against these limits, taking into consideration the historical data of the unit in question.

When a limit is exceeded, an action plan that has been previously agreed and documented as part of an SOP, should be followed. It is normal to have limits at two levels, an 'alert level' and an 'action level'. The 'alert level' is the level of microorganisms which shows a potential drift from those found in normal operating conditions, which, when exceeded, triggers a set of enquiries about the situation. The 'action level' when exceeded initiates a predefined set of actions to remedy the situation.

The organisms cultured which are above the limits should be identified and their likely source determined. Any preventative or remedial action should be identified, agreed, implemented and properly documented. Some organisations follow a policy of three consecutive 'alerts' trigger a remedial action. Limits may not be based on numbers alone and a 'failure' can be based on finding 'undesirable organisms', especially fungi.

9.2 Recommended test protocols

9.2.1 Active air sampling

There are a number of different types of sampler. Each has a method of sampling a known volume of air over a defined period of time. In some samplers, air is drawn over a nutrient agar surface at such a velocity that any particulate contaminants are impacted onto the surface. Impaction can be facilitated by jet velocity or centrifugal force. Other samplers use a gelatine filter. After filtering an air sample, the gelatine may be dissolved in a nutrient medium to culture any organisms trapped by the filter. Alternatively the gelatine filters can be placed onto solid agar surfaces on which any organisms present will grow into visible colonies. Further information on active air sampling is provided in pharmacopoeias and vendor literature.

Samplers should be used according to manufacturers' instructions and recalibrated at the recommended intervals.

Some samplers come in two parts with the suction motor remote from the filter head so that the motor part can be left outside the isolator being sampled. This can be an advantage as one-piece samplers, with their potentially contaminated motors, may not be capable of being decontaminated to an acceptable level. It should be noted that samplers with remote suction motors may not necessarily sample the calibrated volume of air when the sampler head is inside the isolator and the suction motor outside. Positive pressure in the isolator could cause a greater sample volume to be taken and negative pressure could cause a smaller sample volume to be taken. It is therefore recommended that the calibrated volume flow rate is verified before a sampler is used in a given situation.

9.2.2 Settle plates

Before carrying out tests using settle plates please refer to section 9.1 for general guidance.

Settle plates are normally Petri dishes, which always come with their own lids. They should only be opened where and when they are to be exposed. The lids should be replaced whilst the plates are still in the sampling zone and then only subsequently opened after incubation in the testing laboratory. Plates need to be placed in the designated places according to the sampling plan (SOP) (see section 9.1.3). Four hours is suggested as the standard length of exposure for settle plates in an isolator in the non-operational state. Where settle plates are used to monitor contamination during operational sessions, the length of exposure may be shorter but must be accurately recorded for comparative purposes.

Exposure times should not be permitted which allow excessive drying of the plate to take place, as this will reduce the fertility of the media.

9.2.3 Surface sampling of isolator surfaces

This section applies mainly to isolator surfaces, but the techniques can applied to monitoring container surface contamination.

Before carrying out tests on isolator surfaces using contact media please refer to section 9.1 for general guidance.

Ideally, the efficiency of retrieval of surface contamination by swabs or contact media should be known. This information is unlikely to be readily available.

Methods for surface sampling usually use contact plates or swabs. Contact plates are probably more effective and have a greater recovery pickup from a surface than swabs. Swabs have an advantage over contact plates when sampling inaccessible places such as pipes, valves and around equipment.

Surface contact plates, often called 'Rodac' plates are overfilled small Petri dish plates, in which the nutrient media stands proud of the dish, enabling it to be pressed onto the surface to be tested. Other surface contact devices are strips of media on plastic holders presented in sterilised outer sleeves. Both types are used in a similar manner.

The nutrient agar surface must be placed in firm contact with the surface to be sampled at a defined pressure for a defined period of time. Advice should be provided by the supplier of the contact plates or strips. A standard procedure of light hand pressure for 2–5 s is likely to be satisfactory.

After use, the surface should be wiped with a contamination-free swab to remove any nutrient residue left by the media that could later support microbial growth.

Surface swabs are usually sterile absorbent cotton buds on a stick, presented in a sterilised outer tube. The swab is wiped and rotated over a known surface area, usually standardised to 10 cm^2 and returned to the tube. Later the swab pickup is transferred to a nutrient agar for incubation. It is also possible to dilute the swab pickup in a liquid media which is filtered. The filter is then incubated in contact with a semi-solid growth media. This can give a better recovery but it is quite laborious by comparison.

A neutralising media will be necessary for surface samples taken where there may be a residue of disinfecting agent present. This is a specially formulated growth media that counteracts the bactericidal effect of the residual disinfectant. Neutralising media should be used for any sampling which follows immediately after disinfection. This might be, for example, as part of a cleaning validation. It should be noted that an aqueous dilution of alcohol is volatile and will not leave an inhibiting residue, unlike alcoholic solutions of some disinfecting agents.

9.2.4 Finger dabs

Before carrying out finger dab tests please refer to section 9.1 for general guidance.

Finger dab tests are used to monitor contamination of the gloves worn by an operator in a session. The sampling needs to be done at the end of the working session and before any sanitisation of the glove takes place. For both gloves, the pads of the fingers need to be pressed gently but firmly on to the surface of a 90 mm nutrient agar plate. The gloves will need to be wiped free of nutrient material immediately after sampling.

Exposed plates are then incubated and any colonies counted and recorded.

9.2.5 Process validation

Broth fills are used to validate a process. They are designed to simulate the operations carried out by the facility, including interventions during processing that form part of these operations. These interventions might include adjustments to equipment.

The responsible person should agree a suitable procedure that simulates the operations carried out. Consideration should be given to first, middle and last units filled, and the largest batch of material normally handled. The media used for the fill simulation is commonly tryptone soya broth for broad spectrum and thioglycollate broth which, due to its reduced oxygen concentration, allows recovery of both aerobic and anaerobic organisms.

9.2.6 Microbiological validation of the operator

Surface contamination on items transferred into an isolator, occurring when operators do not properly adhere to the procedures, is a possible source of microbial contamination in the controlled workspace and in the product. Therefore validation should be carried out to demonstrate that operators undertaking aseptic processes are doing everything that is necessary to maintain the sterility of the product. In particular:

- gloving procedures should be checked by means of finger dabs;
- surface decontamination by spraying, wiping or dunking should be checked by contact plates in a similar manner to that described in section 9.2.3;
- aseptic techniques should be checked by operator broth fills.

Ways of monitoring the efficiency of surface decontamination are not fully developed. Low initial bioburden levels and possible inconsistencies in technique add to the difficulties of this monitoring. Detection of a significant reduction of surface contamination where the initial bioburden challenge is very low is unlikely to be demonstrated statistically. It is advisable to maintain suitable records that tests have been completed. This monitoring should also reveal deficiencies in the SOP or in the disinfecting or sanitising process itself.

Operator broth fills also form part of process validation (see section 9.2.5).

Further information on validation generally is provided in Chapter 10: Validation.

9.3 Hand-washing facilities

It must be emphasised that hand-washing facilities should not be present in isolator rooms or clean areas. They may be sited outside the clean areas or, if there is no alternative, in the dirty side of the change area.

Regular monitoring should be carried out of any hand-washing facilities in the unit for microbial contamination. It should include regular swabbing tests into drain-traps. This monitoring must be carried out on a regular basis in case circumstances change, even after a series of satisfactory results.

It is suggested that microbial contamination limits for cold water should be based on those applied to potable water. Limits for hot water should be based on local knowledge of the system and recent monitoring history. Current NHS guidelines direct that the temperature of the principal hot water mains is high enough to prevent microbial contamination.

Note that where hot water lines are regularly used and flushed with hot water, there should be no microbial contamination recorded. However, many hot water systems have branches and dead legs in which lower temperature water may harbour viable microorganisms. In critical locations such as aseptic units, any such branches and dead legs should be engineered out if possible, or if not, they should be flushed out with hot water at frequent regular intervals.

9.3.1 Hand washing

Careful training in good hand-washing techniques is essential. Microbial levels should be low.

Monitoring of staff hand contaminants after washing should be carried out at regular intervals. It is generally accepted that a 3-log reduction of contaminating microorganisms is expected by hand washing with soap and water. It is therefore reasonable to aim for at least a 2-log reduction in practice. All organisms should be identified and action taken if the presence of harmful or significant types of contaminants is detected.

It is understood that a group of specialists is working on a British Standard for testing hand-washing products.

The monitoring procedure is similar to glove finger dab testing and is performed directly before and after hand washing. After sampling the hands should be re-washed to remove traces of nutrient material.

9.4 Sterility testing

There should be an accepted local procedure for checking that medicinal products or medical devices, prepared in a pharmaceutical isolator, comply with the British Pharmacopoeia (BP) test for sterility, modified if necessary to permit small batch sizes to be tested. The numbers of items to be tested will need to be calculated from the batch size and the container volume to ensure a reasonable and fair sample is taken for the test. Details of sampling and sample numbers can be found in the relevant pharmacopoeias.

It should be remembered that the sterility test is far from perfect. It may not support growth for all organisms, especially those damaged by cleaning or sanitisation processes. It is recommended that readers be aware of the statistical limitations of sterility testing which should be regarded as only part of an overall quality assurance process.

9.5 Suggested target levels

The recommended maximum acceptable levels of microbial contamination from the different sampling processes are shown in Table 9.3 based on guidance provided by the EC GMP and the Quality Assurance of Aseptic Preparation Services.

9.6 Transfer devices

Some control of microbial contamination levels in transfer devices should be maintained. Although this may be a good aim, it should be noted that every transfer device is different. It is not practical to test inside a type A1 or A2 transfer device. Transfer devices B to E are open to the background environment at some stage; therefore it is not reasonable to expect Grade A standard on surfaces. It would, however, be reasonable to expect Grade A air after a clean-up time. Type F transfer devices should maintain Grade A air and surfaces following sanitisation. Transfer devices operating at negative pressure attached to negative pressure isolators (types C2 and D) draw air from the background

Table 9.3 Recommended maximum acceptable levels of microbial contamination from the different sampling processes

Room air grade	*Suggested target levels of organisms*			
	Finger dabs[a]	*Settle plates (nominally 90 mm)*	*Surface samples*	*Active air sampling*
A (specification for the controlled workspace of isolators)	<1 per plate from 10 digits	1 per 2 plates for up to 4 h sample time[b]	<1 per contact plate or per 10 cm^2 surface sample	<1 per m^3 of air sampled
B	Not applicable	5	5	10
C	Not applicable	50	25	100
D (minimum specification for the background environment for siting isolators)	Not applicable	100	50	200

[a] Clarification of EC GMP guidelines for finger dabs which simply state <1.
[b] This represents a reduced acceptance level for settle plates compared to the EC GMP guidance.

environment when their outer doors are open and so the settle plate results could be adversely affected, depending on how long the outer doors are left open.

Suitable tests and levels, set by the QC microbiologist, should take into account the type of transfer device, the application, the decontamination method and the grade of background environment, if it is considered appropriate. It should be remembered that the ultimate objective of transfers is to get objects into the isolator controlled workplace without contamination. Therefore monitoring and validation of the transfer and decontamination process are more important than monitoring and validation of the transfer device.

9.7 Recommended microbiological sampling frequencies

Table 9.4 lists typical frequencies to be considered for microbiological monitoring in an isolator and its background environment.

Table 9.4 Typical frequencies to be considered for microbiological monitoring in an isolator and its background environment

Interval	*Monitoring activity*
Sessional	• Settle plates – in transfer device and controlled workspaces
	• Finger dabs
Weekly	• Settle plates in transfer devices and work zone (non-operational state)
	• Surface sampling of transfer devices and work zone
	• Settle plates in isolator rooms and change rooms
	• Surface sampling of isolator and change rooms
Quarterly	• Active air sampling of rooms and isolator
	• Spraying in validations (if relevant)
	• Operator broth trials

Note: In addition to the above, the monitoring schedule should include control of any specific design features or identified weaknesses of a particular isolator or its workplace. Attention should be given to inanimate vectors of contamination such as intercom units, buttons on the isolator door handles and other specific design features. If the isolator is serviced by a support room which has a sink, swabs should be taken of the sink.

The above advice is weighted towards hospital aseptic production, but the principles should apply to larger manufacturing units.

10

Validation

In the current climate of quality assurance, regulatory audit and inspections, systems and procedures are formally examined to verify that they do what they are supposed to do. Validation confirms that what we expect to be happening, really is happening.

Scope

All isolator projects will require some form of validation before the equipment can be used. The stages of the validation process are well established and cover every type of isolator. This chapter provides an outline for validation procedures. These will be appropriate both to standard product design isolators as used in hospital pharmacies, and bespoke design isolators or standard form derivatives as used in industry.

10.1 Overview

Validation is defined as establishing documented evidence which provides a high degree of assurance that a specific process will consistently produce a product meeting its predetermined specifications and quality attributes.

Validation is generally against regulatory requirements concerned with pharmaceutical manufacturing, such as GMP (sometimes referred to as cGMP (current good manufacturing practice)). It may also be used to confirm conformance with other relevant directives and health and safety requirements.

Evidence of regulatory compliance and required performance includes reviews of: documentation, instrument calibrations, isolator function and performance testing, together with environmental testing and operator training. Any test result must be compared with acceptance criteria that are based on either sound science or a clear rationale. These criteria should be referenced to published documents, standards, monographs or regulatory requirements.

10.2 Documentation

10.2.1 Document types

Within the process of validation there are a number of different, accepted document types:

- **specification**: defining document giving precise details of a function or a piece of equipment;
- **plan**: document detailing content and schedule for one or more stages of validation;
- **protocol**: document defining scope, overall methodology and rationale, for example for a validation stage. Types and number of qualification tests are included, together with the test progression sequence;
- **procedure**: document defining the specific methodology for a process, activity, check or test, including acceptance criteria;
- **report**: specific or collective, detailed or summary document covering one or more validation stages with results and outcomes, i.e. pass, fail, or conditional pass. The report should include a section on discrepancies with recommendations on all actions to be taken. Every qualification stage will have an interim and a final report. All reports should include a list of supporting documents and their precise location.

In some cases protocols may include procedures.

10.2.2 Document structure

Validation documents must be well structured through a hierarchy. This will ensure that raw data are collected and recorded, and any results, reports and associated actions and recommendations are laid out and cross referenced in such a way as to support audits by quality control and regulatory authorities. Audits may be undertaken by the MHRA (formerly the MCA), FDA (Food and Drug Administration, USA), by local 'internal' or regional 'external' quality control or quality assurance personnel, or by other professional organisations.

Validation documents need to be written taking into account that they will be read in many years' time by people not involved at the start of the project, who will have no guidance other than these documents. The validation documents should state clearly and unambiguously what was done, by whom, why, what was the result, whether the result passed, conformed or failed. If failure occurred, it must state what recommendations were made or what actions taken. This is also referred to as discrepancy reporting.

That a validation process has been well managed and reviewed will be evident by the presence of properly approved documentation, results of testing and sound, well-written reports. A critical factor is a good document storage and retrieval system. The referencing system must be clear and easy to use, allowing any document to be quickly identified and retrieved. The best validation documentation in the world is of no use if it cannot be found when needed.

10.3 Change control

All validation documentation and key defining documents (such as drawings or SOPs) should be subject to a change control system. This will include approval procedures, cross-referencing systems for traceability, with appropriate implementation and recording procedures. Key documents should become 'controlled' documents released only to specific personnel by name or job title with update and revision status control.

Any change should be approved before implementation. The level of approval depends on the impact of the change on the total system. It may be useful to undertake an impact assessment as part of the change review process. This is to verify that the impact of any change is as expected. Tests or checks should be made to ensure that every possible effect of the change is reviewed and the potential outcome recorded.

10.4 Summary of validation stages

The following validation stages or steps are often referred to by their initials:

10.4.1 User requirement specification (URS)

The key defining document against which validation and all qualification to verify compliance is based. The URS represents the user's expectations but should not set limits that may be unnecessarily difficult to achieve or requirements that are not relevant. The clearer and more concise the URS, the better will be the tender documents and the ultimate installation.

10.4.2 Validation master plan (VMP)

A high-level coordinating document giving an overview for the validation of a total system or process comprising individual pieces of equipment or processes. A particular site or unit may already have a VMP in place.

10.4.3 Validation plan (VP)

A specific document describing the validation of an individual piece of equipment or process giving the scope and details of the validation.

10.4.4 Functional design specification (FDS)

Documented details of the manufacturer's design of the equipment to comply with the URS. The key document is the Functional Design Specification (FDS) which describes what the system will do in order to comply with the URS. Additional documents include drawings, detailed specifications, calculations and other data. The FDS should specify tolerances on important dimensions, and performance criteria with alert and action levels. For isolators, it should also include flow/pressure analysis in different operational configurations, e.g. doors open and closed.

10.4.5 Design qualification (DQ)

The documented verification that the proposed design and in particular the FDS complies with the URS and with GMP, and is suitable for the intended purpose. This may involve a check of the manufacturer's calculations and supporting documents, and should ideally be completed before anything is made.

10.4.6 Factory acceptance testing (FAT)

A pre-delivery test of an isolator system which the user or client attends to witness and accept against a written protocol which is normally written by the manufacturer and pre-approved by the user or client. If the FAT protocol is written in such a way that it can be used as the basis for the IQ and OQ protocols, then the IQ and OQ validation stages may be rather easier to carry out.

10.4.7 Type qualification testing (TQT)

A comprehensive set of tests that are completed once for a standard isolator model to establish full compliance with the design as defined above, and relevant standards, before the model is released to the market.

10.4.8 Factory qualification testing (FQT)

A factory test carried out by the manufacturer on every standard model to verify that it complies with its TQT. This is usually less comprehensive than the TQT.

10.4.9 Installation qualification (IQ)

Verification, before the equipment is switched on, that everything is present in accordance with the current drawings and specifications, and that the equipment has been properly installed. IQ also involves checks of the adequacy of documentation. Documentation should at the very least include:

- FAT report;
- certificates of conformity;
- calibration certificates for test instruments used for the FAT;
- calibration certificates for all critical instruments on the equipment;
- operation and maintenance manuals with parts lists, testing schedules and maintenance schedules.

10.4.10 Site acceptance testing (SAT)

Testing on site against a written protocol which is normally written by the manufacturer and pre-approved by the user or client. Testing is to demonstrate compliance with the FDS. The user or client should be in attendance to witness the tests and accept the results.

10.4.11 Operation qualification (OQ)

Verification that the equipment is capable of operating to the FDS. A significant part of OQ comprises examination of all relevant documentation including SAT results. OQ should also include performance tests at the upper and lower operating limits and even, if appropriate, outside the normal operational range. Such tests are sometimes known as 'stressing' tests. Further checks may include safety, ergonomics and the presence of approved standard operating procedures (SOPs).

10.4.12 Performance qualification (PQ)

This is the ultimate series of tests of the design and verifies that the isolator system will meet every requirement of the URS in use. Documentation for these tests should be comprehensive and at the very least include reports on:

- air quality in use;
- microbial contamination in use;
- process output and quality;
- user expectations.

Reliable, robust and consistent performance is critical throughout the lifetime of a piece of equipment and therefore the PQ should be periodically repeated (see section 10.4.7).

10.4.13 Commissioning and validation

It is important not to confuse commissioning with validation. Validation is the whole process, as described in this chapter, of verifying every aspect of a project through a number of formal stages. Commissioning is essentially testing and adjustment on site until the equipment performs as intended. Commissioning can be formalised as site acceptance testing (SAT) to form part of the OQ stage of validation.

10.4.14 Requalification testing

The critical elements of the PQ should be repeated throughout the life of the equipment at defined intervals.

Validation should start at the creation of the system or process and continue right through to decommissioning, i.e. throughout the complete 'life cycle'.

10.5 User requirement specification (URS)

10.5.1 URS bespoke isolators

The URS is typically written by the user or purchaser at the outset of project, possibly with the help of a consultant, or a manufacturer or vendor. The importance of this document should not be underestimated. Validation as a process can only be truly effective if the requirements in terms of quality, function and performance are clearly stated so that test challenges and acceptance criteria have a sound basis against which compliance can be verified. The more precise the description, the less is the risk of subsequent misinterpretation.

The URS is used initially as a document against which bids are invited and tenders adjudicated and subsequently, once the project is awarded, as part of the validation process.

If the URS provided by the client is brief and open to interpretation, the project team should engage with the user to provide a more detailed URS clarifying any ambiguous issues or requirements. Critical to any user requirement is the definition of the process which the isolator is required to protect or contain.

10.5.2 URS standard isolators

Standard isolator requirements should also be derived from a URS. Typically, standard products are 'type tested', that is tested as a product type for a defined use and performance. Product specifications are then produced as a technical data sheet, which can be subsequently referenced as part of a purchase specification, or used as an FDS. The purchase order referencing the technical data sheet can become the URS for the purchase of a 'type tested' product. As with bespoke projects, the URS is used initially as a document against which bids are invited and tenders adjudicated, and subsequently, once the project is awarded, as part of the validation process.

10.5.3 URS content

The URS should include:

- A statement of the process requiring use of isolation technology, such as TPN, cytotoxic reconstitution, CIVAS, BCG (Bacillus Calmette–Guérin), radiopharmacy, sterility testing or other manufacturing processes;
- For standard isolators, a product type or model reference if known, specifying size, configuration, function and materials of construction;
- A statement regarding positive or negative pressure operation;
- Transfer device type and interlock requirements. For more information see Chapter 3: Transfer devices;

- Internal air quality (EC GMP grade) and air flow regime (turbulent or unidirectional);
- Intended background environment;
- Exhaust configuration (external venting or recirculation to the room environment);
- Inlet and exhaust HEPA filtration requirements (specification and double or single);
- Instrumentation requirements, including those that are standard;
- Ergonomic requirements or limitations;
- Cleaning and decontamination methods and requirements, including whether sporicidal gassing is to be used;
- Access and installation requirements and limitations. This should include details of any restrictions that may affect delivery, and details of services to be provided such as electricity, compressed air, vacuum, water, drainage;
- Compliance (regulatory), i.e. GMP, GAMP (good automated manufacturing practice) for automated systems (if applicable) and safety. GMP is sometimes referred to as cGMP, which is simply 'current good manufacturing practice';
- Documentation requirements: these should include operation and maintenance manual with recommended spare parts list, recalibration schedule, TQT certificate (for standard isolators) with certificates of conformity if applicable;
- Validation requirements: while validation is the responsibility of the purchaser, the vendor or manufacturer may be expected to provide IQ/OQ documentation, either to his own format, or to a format specified by the purchaser. This should be clarified in the URS;
- Risk assessment requirements for the isolator as designed in its anticipated application.

Standard isolators may have additional options, including extra features or functions. Any requirements for such options must be clearly stated in the purchase order or URS.

10.6 Validation master plan

As a high level document, the validation master plan (VMP) should state the policy and strategy for total system validation and how different items of equipment and processes are to interact to form a total system. The VMP should list all associated validation documents including individual validation plans and protocols. The list will include those documents in existence and those to be created to complete the validation study. Where existing documents clearly describe procedures and methods, they need only be referred to rather than be reproduced unnecessarily.

Each validation study should ideally have a validation team, led by a validation manager, responsible for the approval of documents, implementation of qualification tests and approval of results. The VMP should define the team and individual responsibilities by title. The team should be kept to a workable number, but ought to include representatives of those departments whose activities impinge on the project, such as quality assurance, engineering and the manager of the user department.

10.6.1 VMP structure and content

1. Introduction and scope This should state the validation policy, an overview of the project, and the scope of the validation activities. The validation personnel will be identified, usually by title and/or department rather than by name, and responsibilities allocated.

2. Description This covers a description of the process and the facilities. Process flow diagrams and facilities layout diagrams are important. The air cleanliness classification is stated, and it is often helpful to give personnel movement diagrams.

3. Documentation This lists the documents to be produced and their format. The approval process for all validation documents will be described. Change procedures must be defined, including methods for alterations to text and drawings. These would be expected to follow accepted conventions. For example, deletions should have a single line drawn through, with initials and date alongside.

4. List of systems/processes to be validated This list defines exactly what is to be validated and what is to be left out. It will also refer to the various SOPs that are pertinent to validation, including change control, maintenance, calibration, cleaning and training.

5. Validation schedule This will detail the timetable for the validation process, including the extent of the validation exercise, the sequence of activities, the number of repeat tests and a clear statement of assumptions.

6. Acceptance criteria These must be carefully defined. All too often a validation exercise will fail because the acceptance criteria are too rigid and cannot be attained. General guidelines for developing acceptance criteria are given in the VMP, although specific criteria are usually given in individual qualification protocols.

7. Discrepancy reporting This will describe how discrepancies will be recorded and subsequently resolved. It is suggested that the details of each discrepancy and its resolution are recorded onto a report form which should then be filed, possibly in the same place as the protocol.

8. Stand-alone isolators In this case a VMP may not be required and a specific validation plan is adequate. A specific validation plan should include a description of the equipment to be validated, a list of required supporting documentation, a schedule of the validation process, and the rationale for acceptance criteria. The format should be broadly similar to a VMP.

10.7 Design qualification (DQ)

The process of design should create key design documents including general arrangement (GA) and installation drawings, pipework and instrumentation diagram (P&ID), functional design specification (FDS), design calculations and any design reviews. DQ verifies and documents that the design, including the FDS, complies with the URS and GMP. The URS, prepared by the user, specifies the user's intentions and may quote international and/or national standards. For isolators this will probably be Draft ISO 14644-Part 7 and key guides, including this one. The FDS, prepared by the manufacturer, specifies the purpose and functions for which the manufacturer has designed the isolator. It is important through the DQ to recognise what elements of the standards or guidance documents quoted in the URS are relevant to the specific application. Regulatory requirements are paramount, but standards are continually developing and may need interpretation for certain applications.

DQ should answer the following questions:

1. Does the design of the isolator comply with the URS and with GMP?
2. Will the isolator be suitable for (capable of) carrying out the tasks for which it has been designed?
3. Will the interfaces with other equipment be functionally effective?

4. Is any specified ancillary equipment suitable for use with the isolator, and is it capable of performing as intended?
5. Is it possible to deliver and install the isolator as designed?
6. Details of equipment control limits – time, pressure, temperature, humidity, airflow.
7. Once installed, can the isolator be operated and maintained as designed?
8. Have there been any changes to the design during the design phase or as the result of a design review, and if so, have they been documented?

A final stage of the DQ may include risk assessment for safety and process failure. A safety assessment can be made by carrying out a formal HAZOP study.

Failure mode and effects analysis (FMEA) can be used as a method of assessing the risk to the process or product posed by a failure in any of the critical areas of the design or operation. One way of carrying out an FMEA is by assigning a score of between 1 (low risk) and 5 (high risk) to each of three aspects:

- the inherent danger presented by the failure;
- the probability of the failure actually occurring;
- the detectability of the failure.

The FMEA value in the critical area is the product of the three values. There is no actual guide as to the score that should trigger further investigation, but a score in excess of 27 (which represents the product of a score of 3 for each value) has been used.

Alternative approaches or methodologies include HACCP and EN1050 Safety of machinery: principles for risk assessment.

10.8 Installation qualification (IQ)

IQ should establish and document that the equipment has been correctly supplied and installed. Information to be collected includes:

1. Equipment identification information – a full description with model and serial numbers;
2. Critical component identification, in particular the filter serial numbers;
3. Verification of correct positioning of isolator and connections to utilities and services, including termination type, rating and position;

4. A schedule of safety functions and features such as alarms and interlocks, electrical protection and safety tests;
5. Supplier documentation – FDS, operation and maintenance manual with wiring diagram, parts lists, testing schedules and maintenance schedules, data sheets, and GA and installation drawings;
6. Materials data sheets and certificates for critical components;
7. A list of all instruments fitted including serial numbers and calibration data and indicating whether they are designated critical or non-critical. Ideally, a data sheet for each instrument should be provided giving the manufacturer, model number, serial number, identification tag number, accuracy required and calibration frequency;
8. Filters, including constructional data and manufacturer's test data. This data should be traceable to the serial numbers;
9. CE marking and other statutory compliance data;
10. Factory acceptance tests (FAT), as reference of manufacturer's correct supply;
11. Software and hardware information relating to any PLC control system. If the software version has been developed for the specific application, it is necessary to have a back-up copy archived. Software should be subject to the same change control conditions as any other documents. Detailed guidance on the validation of software is given in GAMP which is described briefly in section 10.11.

10.9 Operational qualification (OQ)

The OQ is a crucial stage, and there is also a preferred format for the OQ protocol as follows:

1. The cover page includes the title of the study, the identification and location of the equipment, and the unique document reference;
2. An approval page to indicate that named responsible persons have seen and have approved the protocol prior to any work being undertaken. This page may also include a section to show that these same named responsible persons have reviewed the results and have approved them;
3. The responsibilities of the named responsible persons who will carry out the OQ;
4. A summary of the results of the OQ with indication of pass or fail;
5. An overview of the purpose of the OQ with reference to related documentation;

6. A full description of the equipment and what tests are to be carried out, including SAT and any additional stressing tests. Stressing tests are tests outside the normal operating range. Further checks may include checks regarding safety, ergonomics, the presence of approved SOPs and other site-specific requirements. There should be a statement about how non-compliance will be treated, and a time schedule for the work. This is especially important for project management purposes, otherwise the testing might never finish. The procedure for each test should describe the test and provide details of the acceptance criteria and acceptable limits of deviation.
7. A detailed table of results, with pass or fail statement, and allowing for dated signatures of the test engineer, and possibly a witness. The actions taken as a result of failure may also be shown here. This should be supported by the data listed in items 10–15 below.
8. Serial numbers and calibration certificates for all test instruments used and reference to the validation of the test method if applicable;
9. Supporting documentation. This should include a comprehensive review of SAT results and a check on the availability of documents such as equipment logbooks, maintenance programmes, SOPs and operator training schemes. These last two may be written during the OQ. See also item 16. There should normally be SOPs for gowning in preparation for the use of the isolator, cleaning and decontamination of the isolator, the operation, use, monitoring and care of the isolator, cleaning and decontamination of the components entering the isolator, all activities and processes to be carried out in the isolator, dealing with spillages, breaches and other mishaps, and the use of ancillary equipment.

The OQ protocol must show that the equipment can operate according to the URS and manufacturer's specifications. Information supplied must be accurate and include:

10. Data demonstrating that the equipment operates within its control limits – time, pressure, temperature, humidity, airflow;
11. Data demonstrating satisfactory function to the design performance including in situ testing of HEPA and ULPA filters, leak tests, process transfer, interlock and alarm function tests (including compliance with safety instructions or guidelines) and specific requirements such as breach velocity tests for containment isolators;
12. Data to show consistency and reliability of equipment operation, including air distribution;

13. Background and internal controlled workspace environmental data, including viable and non-viable particle monitoring;
14. Data to demonstrate software checks;
15. Qualification of sporicidal gassing cycles, including temperature mapping, gas distribution and gas concentration studies, see Chapter 6: Cleaning, decontamination and disinfection;
16. Approval of content of supporting documentation listed in item 9.

Isolator operator training, an essential requirement for performance qualification, should start as early as is practicable. Ideally, training should start immediately following OQ, when the system is known to operate effectively and safely and when the SOPs have been written. Advice on training is given in Appendix 2: Training.

10.10 Performance qualification (PQ)

PQ should not start until the OQ has been completed and signed off, and any critical issues resolved. PQ should demonstrate and document that the properly designed, installed and operating isolator can, with reliability and consistency, be used for the purpose for which it was designed. The information to be collected should include:

1. Data showing that the equipment performs within the limits established in the design stage through a sufficient number of repeat cycles to demonstrate consistency and reliability. The data should include internal pressure, airflow, temperature, humidity, particulate levels, and microbial contamination;
2. Data showing that the cleaning and decontamination procedures for the controlled workspace of the isolator and components entering the isolator produce the designed reduction in surface contamination, and that the cleaning residues are within acceptable levels;
3. For gassed isolators, data showing that the developed gassing cycle will reliably and reproducibly result in a predetermined reduction in bioburden on the surfaces of the controlled workspace of the isolator and components entering the isolator, verified by inactivation of biological indicators of defined type and mean spore population. Gassing cycles are normally developed during PQ. Because of the potential interactions of sporicidal gases with the product, gassing times should be validated with respect to the effect on product quality;

4. Data showing that the defined operations can be safely, reliably and reproducibly performed. This may involve broth trials or other process simulation;
5. Data showing that the desired interaction with other equipment complies with the specifications, including the isolator ancillary items such as transfer devices.

SOPs can be refined and validated during PQ.

Some isolators may have associated processes or ancillary equipment with separate DQ/IQ/OQ validation documents. For PQ, these should be combined with those for the isolator system itself. A good example is a sporicidal gassing system. PQ also requires the validation of the interfaces and process transfers between separately validated equipment connected to the isolator, for example autoclaves or freeze dryers (lyophilisers).

For more information on sporicidal gassing cycle development and qualification together with cleaning validation see Chapter 6.

10.11 GAMP – Good automated manufacturing practice

GAMP is a guide that has been prepared under the auspices of ISPE (International Society of Pharmaceutical Engineers), overseen by the GAMP Forum Steering Committee and assisted by a large number of individuals from many of the most important pharmaceutical companies worldwide. The current edition of the guide is GAMP 4, which was published by ISPE at the end of 2001. The purpose of the guide is to 'assist companies in the healthcare industries, including pharmaceutical, biotechnology and medical device, to achieve validated and compliant automated systems'. Automated equipment includes 'standard, configurable, and customisable products, as well as custom (bespoke) applications'. Another quote from the scope reads: 'The automated system consists of the hardware, software, and network components, together with the controlled functions and associated documentation. Automated systems are sometimes referred to as computerized systems'.

GAMP 4 sets out in detail, with flowcharts and useful examples of documentation, a methodology for validation that is entirely consistent with the rest of this chapter including the terminology. Indeed it is possible to use the format of the documentation in GAMP 4 for validation generally and not just for the control elements. This would achieve consistency in the validation documentation of a complete project including its control system.

For software itself, GAMP 4 sets out the validation approach to the different categories or types of software, namely: operating system, firmware, standard software packages, configurable software packages or custom (bespoke) software. Appendix O9 of GAMP 4 aligns GAMP with Annex 11 of the EC GMP – Computerised Systems.

It is recommended that everyone involved in validation should familiarise themselves with the current version of GAMP. Validation of isolator control systems should be the responsibility of those who are qualified to work to GAMP. The whole of this chapter is likely to be a helpful introduction to the principles and terminology of validation for those about to look at GAMP for the first time.

10.12 Summary

The main stages of isolator validation are summarised in Figures 10.1 and 10.2.

The first is for standard isolators that are frequently found in hospital pharmacies, and the second is for bespoke isolators that are mostly found in industry. The information is simplified, and reference should be made to the main text for a full description.

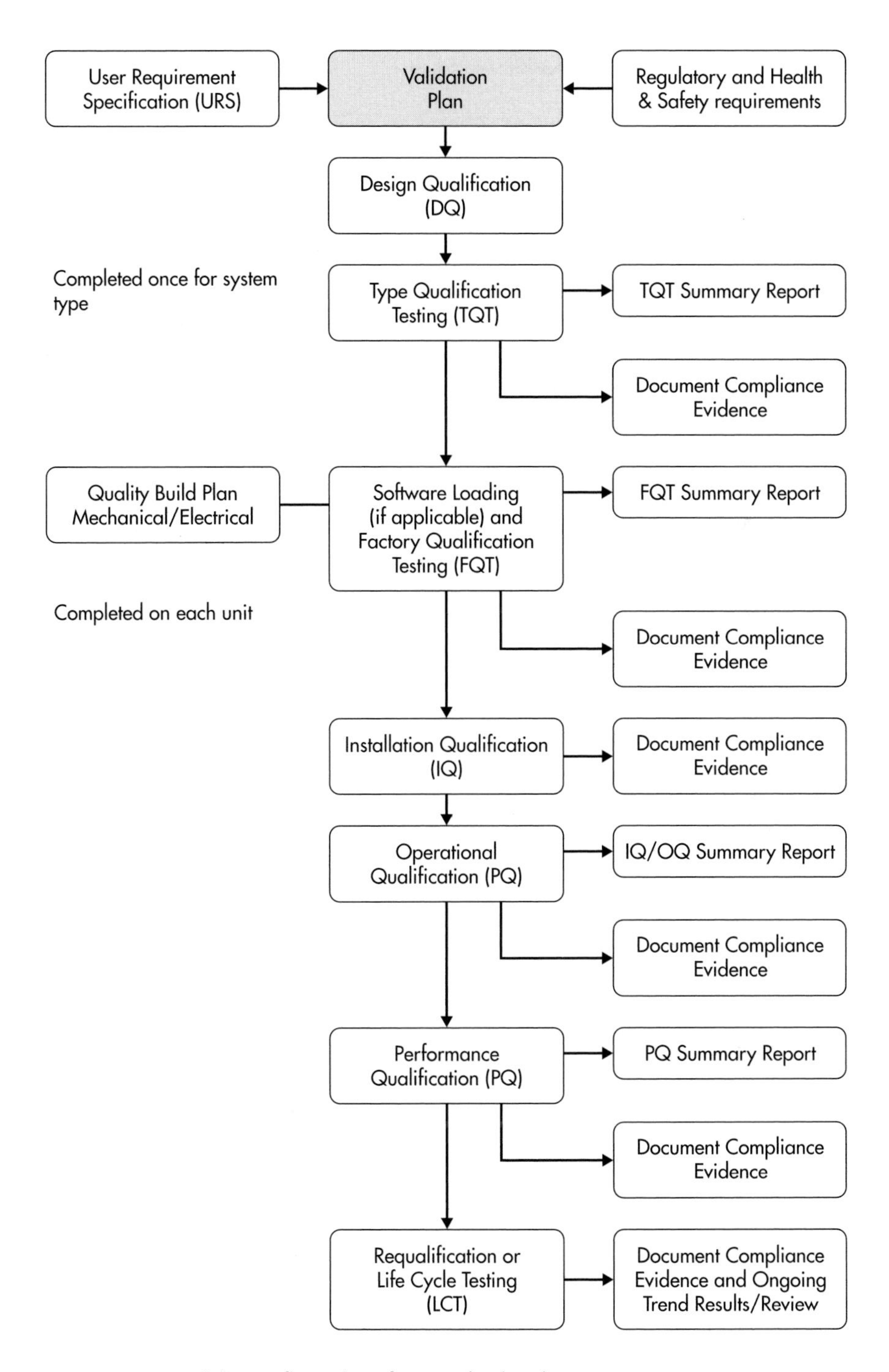

Figure 10.1 Validation flow chart for standard isolators.

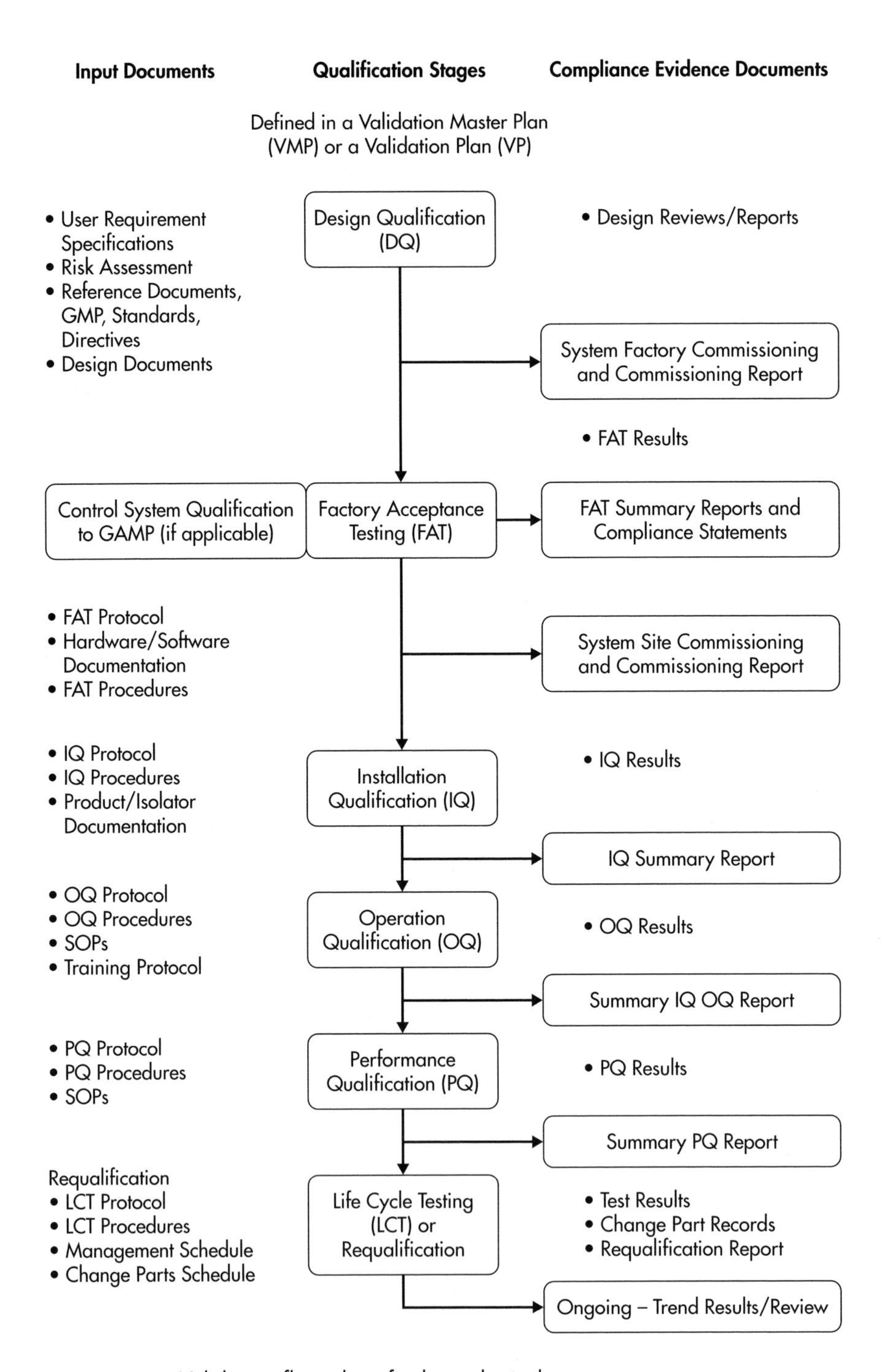

Figure 10.2 Validation flow chart for bespoke isolators.

11

Standards and guidelines

With so many standards in existence it may be difficult to keep track of what is current, what has been superseded, what is developing and what relates to pharmaceutical isolators. This section provides an appraisal of existing and developing standards and explains how to interpret the stages of new standard development.

Introduction

This chapter lists and summarises those standards and guidelines, current or in preparation, that are directly or indirectly relevant to the specification, design, siting, validation and operation of pharmaceutical isolators. It also lists some of the familiar national standards that have now been superseded by the new European (EN) and international (ISO) standards. Many such new documents have been published in the past year or two, and there are more to come. This chapter can therefore only provide a snapshot of what has been published at the time of writing – in final form or in draft. The list of standards and guidelines is not exhaustive.

11.1 Designation and terminology

Readers should be aware of standards designations.

- **BS** is the prefix used for British Standards prepared or adopted by the BSI.
- **EN** is the prefix that is used for European Standards prepared or adopted by CEN.
- **CEN** stands for European Committee for Standardization (or Normalisation in French).
- **ISO** is the prefix for international standards prepared or adopted by the International Standards Organization (ISO).

Thus BS EN ISO can designate a standard developed by the ISO, adopted as a European Standard, and therefore automatically adopted as a British

Standard, or it can designate a standard developed by CEN and adopted both as an ISO Standard and, automatically, as a British Standard.

ISO Standards, especially those that predate the development of European Standards, are not necessarily adopted as European Standards. ISO 10648: Containment enclosures, is an example. However, when a new European (CEN) Standard is published, CEN members are required to withdraw any conflicting or overlapping national standard(s). There is no obligation to withdraw a national standard on the publication of an ISO Standard. Thus, with the publication of the various parts of BS EN ISO 14644-1, BS 5295-Part 1 has been withdrawn.

ISO Standards and parts of standards are prepared by Working Groups, made up of technical experts appointed by national standards committees, reporting to the ISO Technical Committee (TC) or Sub-Committee (SC) of the Technical Committee for that standard. A TC or an SC is made up of the national standards bodies of the ISO member nations that have participating or observing status on that committee. The process of preparing an ISO Standard has a number of stages:

- **WD** Working Draft – this is the draft for seeking the consensus of the members of a Working Group.
- **CD** Committee Draft – this is the draft that represents the initially agreed output of the Working Group. Consensus is then sought within the TC or SC for the draft to go forward to the next stage, which is the DIS.
- **DIS** Draft International Standard – this is the draft that is issued to all participating and observing national standards bodies and other interested parties and is the final stage at which technical comments can be made. These comments are collated nationally and accompany the vote to the ISO central secretariat. A successful ballot allows the document to progress to the final, FDIS, stage. The DIS stage is commonly recognised as the stage for public comment.
- **FDIS** Final Draft International Standard – this is the final draft for issue to all participating or observing standards bodies for final voting. TC or SC members can vote in favour of accepting the FDIS for publication as a standard with only editorial comments being made. If a member votes negatively at this stage, technical reasons must accompany the vote.

The final published version of the standard is recognised because it carries none of these designations, but does show the year of publication.

For European Standards, the designation is different. A draft standard is designated prEN with the pr being an abbreviation of provisional. When the final standard is published, the pr is dropped. Balloting on the prEN document is the final stage at which technical comments can be made by the CEN members. Following this, a formal vote is held which is similar to the FDIS voting within ISO. It should be noted that Standard compliance is a legal requirement in certain European countries but not necessarily in the UK.

Standards are written in a convention that distinguishes between mandatory and non-mandatory requirements. Table 11.1 sets out the terminology normally used.

Non-mandatory requirements should still be assumed to carry considerable weight, as they are likely to represent best practice.

Table 11.1 Terminology used in standards

Mandatory	*Non-mandatory*
Normative	Informative
Shall	Should
Must	May

11.2 Clean air and isolator standards

The principal standard that will eventually cover most aspects of both clean air and isolators is BS EN ISO 14644: Cleanrooms and associated controlled environments. This is being prepared by TC 209 of ISO and has eight parts. Part 7: Separative devices (clean air hoods, glove boxes, isolators and mini-environments), will be the part that deals with isolators. ISO 14698: Cleanrooms and associated controlled environments – Biocontamination control, is being prepared by the same TC of ISO.

Table 11.2 lists most of the principal standards for clean air and isolators, and includes some that are in the process of being withdrawn.

11.3 Filter standards

BS EN 1822: High efficiency air filters (HEPA and ULPA), is the 'new' European filter standard. Part 1 classifies filters according to efficiencies based on the penetration at the MPPS, most penetrating particle size. There are a number of different mechanisms of filtration at work in a HEPA filter. The combined effect of these is that there is one particular size of particle, the MPPS, for which the penetration is at its highest (and the efficiency at its lowest). For particle sizes less than, as well as greater than, this size, the penetration decreases again (and the efficiency increases). It is incorrect to assume that a filter that is, say, 99.997% efficient at 0.5 µm has a lesser efficiency with smaller particles. See also Appendix 4: HEPA Filtration mechanisms, and Table 11.3.

11.4 Biotechnology Standards

BS EN 12469, the new European Standard for microbiological safety cabinets was 'developed by CEN/TC 233 *Biotechnology* in support of 90/219/EEC Council Directive of 23 April 1990 on the contained use of genetically modified organisms, and 90/679/EEC Council Directive of 29 November 1990 on the protection of workers from the risks related to exposure to biological agents at work. It also applies to microbiological safety cabinets used for other purposes including containment of biological agents and infected animals'.

Because one of these Directives relates to genetically modified organisms (GMOs), it is very likely that isolators for gene therapy products will come within the scope of these Biotechnology Standards as well as GMP.

BS EN 12469, the other European Biotechnology Standards that support it, and the old BS 5726 which has been partially withdrawn, are listed in Table 11.4.

11.5 Other relevant standards

The standards in Table 11.5 are also relevant to pharmaceutical isolators.

11.6 GMP guidelines

11.6.1 EC GMP 2002 and Orange Guide 2002

The rules (pharmaceutical legislation) governing medicinal products in the European Union are published in nine volumes. The volume that concerns pharmaceutical isolators is Volume 4 'Medicinal products for human and veterinary use: Good manufacturing practices', commonly referred to as the EC GMP. A new 2002 edition of the EC GMP has been published and is also available in the UK as the Orange Guide, which is simply the UK MCA (MHRA) republication of the EC GMP in the traditional orange covers. More recently in May 2003, a revised 'Annex 1: Manufacture of sterile medicinal products' was published for implementation in September 2003.

The 2002 EC GMP includes four additional Annexes:

- Annex 15. Qualification and validation (July 2001)
- Annex 16. Certification by a Qualified Person and Batch Release (July 2001)
- Annex 17. Parametric Release (July 2001)
- Annex 18. Good manufacturing practice for active pharmaceutical ingredients (July 2001)

The revised 2003 Annex 1 contains the following statements:

> 'A unidirectional air flow and lower velocities may be used in closed isolators and glove boxes.'

> 'In general the area inside the isolator is the local zone for high risk manipulations, although it is recognised that laminar air flow may not exist in the working zone of all such devices.'

It should also be noted that the revised maximum permitted number of 5-μm particles/m^3 in Grades A (at rest and in operation) and B (at rest), which is one, is impractical to test due to the excessively long sampling time to count 20 particles as required in BS EN ISO 14644-1. Furthermore, although the maximum permitted number of 0.5-μm particles/m^3 corresponds approximately to the number specified in ISO 5, this is not the case for 5-μm particles. Revised Annex 1 specifies one particle and ISO 5 specifies 29 particles. The values expressed for ISO classes are derived from the formula in BS EN ISO 14644-1 and should be considered to be realistic.

Subject to the anomaly with 5-μm particles, Annex 1 sets out a very helpful classification of clean environments. The classification gives an 'at rest' and an 'in operation' particle count requirement for the four grades of air from A to D, 'at rest' being the condition that should be achieved within 15 to 20 min after operations have ceased. Confirmation of this clean-up capability should be of far more interest to the user than confirmation of the 'at rest' condition itself. 'In operation' airborne and surface microbial limits are also specified.

Some of the other annexes are also relevant to pharmaceutical isolators, notably:

- Annex 2: Manufacture of biological medicinal products for human use
- Annex 3: Manufacture of radiopharmaceuticals
- Annex 15: Qualification and validation

The EC GMP is compulsory reading for anyone seriously involved in aseptic work and has the great merit of being available free of charge on the internet. An easy way to reach it is by searching for 'Eudralex' on an internet search engine, and then Volume 4.

11.6.2 Quality Assurance of Aseptic Preparation Services 2000

This is a guidance book on aseptic preparation services produced by the NHS Quality Assurance of Aseptic Preparation Services Working Group (Beaney 2000, ref. 2 in Appendix 2, p. 208).

11.6.3 A Code of Practice for Tissue Banks providing tissues of human origin for therapeutic purposes (DoH 2001)

The title is self explanatory and the text can be found at www.doh.gov.uk/humantissuebanking/tissuebank.pdf

11.6.4 GAMP Guide for Validation of Automated Systems

GAMP is a guide that has been prepared under the auspices of ISPE (International Society of Pharmaceutical Engineers). The current edition, GAMP 4, was published in 2001 and is very comprehensive. Although the title indicates that it applies to automated systems, it is written in a format that can be applied generally to validation of equipment in the healthcare industries, including pharmaceutical, biotechnology and medical devices. Further information on how to obtain the guide may be found on the ISPE website: www.ispe.org.

11.6.5 PIC/S Recommendation PI 014-1 24 June 2002: Isolators used for aseptic processing and sterility testing

PIC/S is the Pharmaceutical Inspection Cooperation Scheme and is a convention of regulators from 26 different countries mainly from Europe, but also including Australia, Canada, Malaysia and Singapore. The USA is not a member.

The PIC/S Recommendation referred to here addresses only 'isolators that are subject to a sporicidal process (usually delivered by gassing)'. The Recommendation is one of a number that set out what an inspector should look for when he audits an installation or an activity. These Recommendations all therefore give a useful insight into the approaches of regulatory auditors.

PICS has an extremely good website, www.pischeme.org, from which free downloads are possible.

11.6.6 PDA Technical Report No. 34: Design and validation of isolator systems for the manufacturing and testing of health care products

This technical report, which is one of a series published by the American Parenteral Drug Association, sets out the American perspective on isolators, which differs in a number of respects from that in Europe.

11.6.7 FDA Sterile drug products produced by aseptic processing (Draft)

This was issued in September 2002 as a 'Preliminary Concept Paper not for Implementation'. Appendix 1 is 'Aseptic processing isolators'. The acronym cGMP, widely used by Americans, is used in this document but not defined. In Europe, GMP is that which is laid out in the EC GMP. cGMP seems to be altogether more of a moving target.

11.7 Biotechnology and safety guidelines: ACDP (Advisory Committee on Dangerous Pathogens) documents

The ACDP is appointed by the UK Health and Safety Commission as part of its formal advisory structure. ACDP publications are available from HSE Books (PO Box 1999, Sudbury, Suffolk CO10 2WA. Tel: 01787 881165; http://www.hsebooks.co.uk/books)

11.7.1 Categorisation of biological agents according to hazard and categories of containment (Fourth edition, 1995)

This has been withdrawn and is being replaced by 11.7.2 to 11.7.5.

11.7.2 Second supplement to: Categorisation of biological agents according to hazard and categories of containment (Fourth edition, 1995)

This is an interim document available on the HSE website which includes an up to date list of biological agents.

11.7.3 The management, design and operation of microbiological containment laboratories

This is the first of three parts of the document that is replacing 'Categorisation of biological agents according to hazard and categories of containment (Fourth edition, 1995)' and was published by HSE Books in 2001. Appendix 6 'Microbiological safety cabinets' seeks to tighten up, in the UK, aspects that the HSE consider inadequate in BS EN 12469. These aspects include double filtration on cabinets that discharge to the room and routine testing. It also introduces the concept of 'in-use operator protection factor testing'.

11.7.4 Biological agents – managing the risks

This will be the second of three parts of the document that is replacing 'Categorisation of biological agents according to hazard and categories of containment (Fourth edition, 1995)' and will be available on the internet when complete. It will contain the list of biological agents and, when it is published, the 'Second supplement' mentioned in section 11.7.2 will be withdrawn.

11.7.5 Working with hazard group 4 agents

This will be the third of three parts of the document that is replacing 'Categorisation of biological agents according to hazard and categories of containment (Fourth edition, 1995)' and will be available on the internet when complete. The title is self-explanatory.

11.8 COSSH

Designers and users of pharmaceutical isolators should take into account COSHH regulations.

The current legal base is that 'The principal legislation which applies is the Control of Substances Hazardous to Health (COSHH) Regulations 2002 (S.I. 2002/2677). COSHH requires employers to weigh up the risks to the health of their employees arising from exposure to hazardous substances and to prevent, or where this is not reasonably practicable, adequately control exposure. Employers might also have to monitor employees' exposure to a hazardous substance and place them under health surveillance'.

11.8.1 EH40 Occupational exposure limits

This contains a list of maximum exposure limits and occupational exposure standards for use with the COSSH Regulations 1999 and is updated annually.

Publications relating to COSSH are available from HSE Books.

Table 11.2 Most of the principal standards for clean air, isolators and filters

Standard	*Title and description*	*Status*
BS EN ISO 14644-1:1999	Cleanrooms and associated controlled environments-Part 1: Classification of air cleanliness. *This part supersedes BS 5295 Parts 1 and 4. A formula and table give a completely new classification of air cleanliness covering a range of particle sizes from 0.1 μm – 5.0 μm. The nearest equivalent of old Class 100 (to Federal Standard 209E) is ISO Class 5, and old Class 10 000 is ISO Class 7. Ultrafine particles <0.1 μm are defined by U-descriptors and macroparticles >5.0 μm are defined by M-descriptors. The designation of air cleanliness should specify the occupancy state (as-built, at rest or operational) and the particle size or sizes. Compliance must be demonstrated with a UCL of 95%. The following Annexes are included in Part 1:* Annex A Graphical illustration of classes Annex B Determination of particulate cleanliness using a discrete-particle-counting, light-scattering instrument *Subheadings include:* *Establishment of sampling locations* *Sampling procedure* *Recording of results* *Requirement for computing 95% UCL* *Interpretation of results* Annex C Statistical treatment to derive 95% UCL Annex D Worked examples Annex E Considerations for ultrafine and macroparticles Annex F Sequential sampling procedure	Published.

Table 11.2 *Continued*

Standard	*Title and description*	*Status*
BS EN ISO 14644-2:2000	Cleanrooms and associated controlled environments-Part 2: Specifications for testing and monitoring to prove continued compliance with BS EN ISO 14644-1 *This part supersedes BS 5295-Part 4 and deals with frequency of monitoring only. Monitoring data is used to determine the frequency of the specified test (14644-1 Annex B) to demonstrate particle compliance. Additional tests for airflow velocity, airflow volume and air pressure are at a maximum 12-monthly interval. The following Annexes are included in Part 2:* Annex A Optional tests *These are listed as:* *Installed filter leakage* *Airflow visualisation* *Recovery* *Containment* *Leakage* *and all have a suggested maximum time interval of 24 months. The cross-references to the relevant test procedure in Part 3 are also given.* Annex B Guidance on influence of risk assessment on cleanroom or clean zone tests and monitoring *The content of this annex is not much longer than the title!*	Published.
Draft ISO 14644-3	Part 3: Metrology and test methods *This part will contain a table of recommended tests for IQ with a brief description of the purpose of each test. The list of tests will include:* *Airborne particle counts* *Airflow tests* *Air pressure difference test* *Installed filter system leakage test** *Recovery test* *Containment leak test*	At the DIS (Draft International Standard) stage. Changes may be required and publication could be as late as 2005.

Table 11.2 *Continued*

Standard	*Title and description*	*Status*
	Annex A *will set out the choice of recommended tests and the sequence in which to carry them out.* Annex B *will set out all the test procedures.* Annex C *will describe all the measuring instruments.* **The pass/fail criteria of 0.01% for scan tests* and *volumetric tests are more relaxed than they were in BS 5295.*	
BS EN ISO 14644-4:2001	Cleanrooms and associated controlled environments – Part 4: Design, construction and start-up *This sets out:* *Specification of requirements* *Planning procedures* *Construction and start-up* *Testing* *Approval* *Documentation.* *Annexes expand on:* *Control and segregation concepts* *Development and approval (qualification)* *Layout* *Construction and materials* *Environmental control* *Control of air cleanliness* *Additional specification of requirements.*	Published.

Table 11.2 *Continued*

Standard	*Title and description*	*Status*
Draft ISO 14644-5	Part 5: Operations *This will cover:* *Operational systems* *Cleanroom clothing* *Personnel* *Stationary equipment* *Materials* *Portable and mobile equipment* *Cleanroom cleaning.* *There will be informative annexes.*	FDIS issued. Final publication should be in 2004.
Draft ISO 14644-6	Part 6: Terms and definitions *This will follow when all the other parts have been completed*	This will be the last part to be published.
Draft ISO 14644-7	Part 7: Separative devices (clean air hoods, gloveboxes, isolators, minienvironments) *This is the part that will cover isolators, but it will not be application specific. It will describe a 'Continuum' of aerodynamic and physical means of separation, a range of access devices and a range of transfer* devices derived *from 'Isolators for Pharmaceutical Applications' (HMSO 1994). One of the Annexes will cover Leak testing with reference to ISO 10648: Containment enclosures: Part 2. This Annex will give a mass balance formula for estimating appropriate leak rates.*	FDIS in preparation.
Draft ISO 14644-8	Part 8: Molecular contamination *This is in the early stages.*	DIS in preparation.

Table 11.2 *Continued*

Standard	*Title and description*	*Status*
Draft ISO 14698	Cleanrooms and associated controlled environments – Biocontamination control *This series of standards is being developed alongside the 14644 series. It will contain three parts but the third part has been relegated to become a Technical Report rather than a full standard):* Part 1: General principles Part 2: Evaluation and interpretation of biocontamination data Part 3: Methodology for measuring the efficiency of processes of cleaning and (or) disinfection of inert surfaces bearing biocontaminated wet soiling or biofilms	Parts 1 and 2 have been published. Part 3 has been relegated to the status of a TR (Technical Report) and will be edited when time allows.
Federal standard 209E	Airborne particulate cleanliness classes in cleanrooms and clean zones *This is the fifth and final edition of what has been the father of all clean air standards. The imperial classification, which is stated alongside its metric equivalent, is so easy to visualise. Class 100 means that there are no more than 100 particles greater than 0.5 μm in one cubic foot of air! 209E has been superseded by BS EN ISO 14644-Parts 1 and 2.*	Withdrawn.
BS 5295:1989	Environmental cleanliness in enclosed spaces Part 1: Specification for clean rooms and clean air devices Part 2: Method for specifying the design, construction and commissioning of clean rooms and clean air devices (Incorporating Amendments Nos. 1 and 2 – March 2000) Part 4: Specification for monitoring clean rooms and clean air devices to prove continued compliance with BS 5295-Part 1	Parts 1, 2 and 4 have been withdrawn following publication of the corresponding parts of BS EN 14644.

Table 11.2 *Continued*

Standard	*Title and description*	*Status*
BS 5295-0:1989	Part 0: General introduction, terms and definitions for clean rooms and clean air devices (Incorporating Amendment No. 1 – March 2000)	Amended to align with BS EN ISO 14644. Will be withdrawn on publication of BS EN ISO 14644-6.
BS 5295-3:1989	Part 3: Guide to operational procedures and disciplines applicable to clean rooms and clean air devices (Incorporating Amendment No. 1 – March 2000)	Amended to align with BS EN ISO 14644. Will be withdrawn on publication of BS EN ISO 14644-5.
PD 6609:2000	Environmental cleanliness in enclosed spaces – Guide to test methods *This PD (Published Document) reproduces test methods, including* in situ *filter leak tests, from BS 5295-1:1989 which was withdrawn upon publication of BS EN ISO 14644-1:1999.*	This is an interim document pending publication of BS EN ISO 14644-3, which will cover test methods in full.
ENV 1631 July 1996	Cleanroom technology – Design, construction and operation of cleanrooms and clean air devices	This has been withdrawn following publication of Draft ISO 14644-5 for comment.

Table 11.2 *Continued*

Standard	*Title and description*	*Status*
ISO 10648-1	Containment enclosures – Part 1: Design principles *This standard was prepared by Technical Committee ISO/TC 85, Nuclear energy, Subcommittee SC 2, Radiation protection. The scope, which is considerably wider than just for the nuclear field, starts as follows: 'This part of ISO 10648 applies to enclosures or enclosure lines intended to be used for work on:* *– radioactive and/or toxic products where containment is required for protection of personnel and environment,* *– sensitive products requiring a special atmosphere and/or a sterile medium'.* *The content, however, consists of a mass of constructional detail, much of it not relevant to pharmaceutical isolators.*	ISO standard not adopted by BS/EN.
ISO 10648-2	Containment enclosures – Part 2: Classification according to leak tightness and associated checking methods *This standard, which was prepared by the same committee that prepared Part 1, gives a classification for leaktightness and describes very clearly some useful methods for testing leaktightness in the factory, before commissioning on site and during operation. These methods include the oxygen method, the pressure change method and the constant pressure method. This part of ISO 10648 is referenced in Draft ISO 14644-Part 7.*	ISO standard not adopted by BS/EN but referenced in Draft ISO 14644-7.
AS 4273-1999 + Amdt 1/2000.05.2	Design installation and use of pharmaceutical isolators *This Australian standard, the previous edition of which was published in 1995, is based upon and very similar to 'Isolators for Pharmaceutical Applications' (HMSO 1994).*	Published.
BS 7989:2001	Specification for recirculatory filtration fume cupboards *This specifies safety requirements, carbon filtration systems, as well as the methods used for efficiency. It also stipulates certain levels of performance to be achieved by testing the fume cupboards.* *BS 7989:2001 was produced in response to the industry's request for a standard for fume cupboards of the filtration type, not covered by BS 7258*	Published.

Table 11.2 *Continued*

Standard	*Title and description*	*Status*
prEN 13824	Sterilisation of medical devices – Validation and routine control of aseptic processes – Requirements and guidance *This standard applies to sterile liquid medical devices and their containers which are not terminally sterilised.*	Under approval.
ISO 13408	Aseptic processing of healthcare products Part 1: General requirements Part 2: Filtration Part 6: isolator and barrier technologies.	Part 1 published (1998) but under revision. Part 2 has been published. Part 6 is at the CD stage.

Note: UCL, upper confidence limit.

Table 11.3 Filter standards

Standard	*Title and description*	*Status*
BS EN 1822-1:1998	High efficiency air filters (HEPA and ULPA) – Part 1: Classification, performance testing, marking *This standard, which is for the manufacturers of filters, describes the concept of 'most penetrating particle size' and gives a new classification table for HEPA and ULPA filters. In addition, it is advisable to specify the design flow rate and the proposed* in situ *test method.*	Published.
BS EN 1822-2:1998	High efficiency air filters (HEPA and ULPA) – Part 2: Aerosol production, measuring equipment, particle counting statistics	Published.
BS EN 1822-3:1998	High efficiency air filters (HEPA and ULPA) – Part 3: Testing flat sheet filter media	Published.
BS EN 1822-4:2000	High efficiency air filters (HEPA and ULPA) – Part 4: Determining leakage of filter elements (Scan method)	Published.
BS EN 1822-5:2000	High efficiency air filters (HEPA and ULPA) – Part 5: Determining the efficiency of the filter element	Published.
BS 3928:1969	Method for sodium flame test for air filters (other than for air supply to I.C. engines and compressors)	Published, current and still used by some manufacturers of HEPA filters.

Table 11.4 Biotechnology standards

Standard	*Title and description*	*Status*
BS EN 12469:2000	Biotechnology – Performance criteria for microbiological safety cabinets *This supersedes the mandatory parts of BS 5726. The focus is on cleanability, sterilisability and leaktightness with respect to microorganisms as per the standards that are listed below. It is very likely that in the UK the HSE, via the ACDP (Advisory Committee for Dangerous Pathogens), will be publishing additional mandatory requirements including for on-site operator protection testing at installation and routine maintenance.*	Published.
BS EN 12296:1998	Biotechnology – Guidance on testing procedures for cleanability	Published.
BS EN 12297:1998	Biotechnology – Guidance on testing procedures for sterilizability	Published.
BS EN 12298:1998	Biotechnology – Guidance on testing procedures for leaktightness	Published.
BS EN 12128:1998	Biotechnology – Laboratories for research, development and analysis – Containment levels of microbiology laboratories, areas of risk, localities and physical requirements *This standard includes within its scope the handling of genetically modified micro-organisms and may therefore be relevant where gene therapy products are handled.*	Published.
PD 6632:1998/ CR 12739:1998	Biotechnology – Laboratories for research, development and analysis – Guidance on the selection of equipment needed for biotechnology laboratories according to the degree of hazard	This PD (Published Document), is exactly the same as CR 12739:1998 which is a full European Standard. It has the status of a British Standard.

Table 11.4 *Continued*

Standard	*Title and description*	*Status*
BS 5726:1992	Part 1: Specification for design, construction and performance prior to installation	Withdrawn following publication of BS EN 12469:2000.
BS 5726:1992	Part 2: Recommendations for information to be exchanged between purchaser, vendor and installer and recommendations for installation	This part, which contains the very helpful diagrams on siting, is retained but is being rewritten.
BS 5726:1992	Part 3: Specification for performance after installation	Withdrawn following publication of BS EN 12469:2000.
BS 5726:1992	Part 4: Recommendations for selection, use and maintenance	This part is retained but is being rewritten.

Table 11.5 Other standards relevant to pharmaceutical isolators

Standard	*Title and description*	*Status*
BS EN ISO 14937:2001	Sterilization of health care products. General requirements for characterization of a sterilizing agent and the development, validation and routine control of a sterilization process for medical devices.	Published.
ISO/TS 11139	Sterilization of healthcare products: Vocabulary	Published.
BS EN 1276:1997	Chemical disinfectants and antiseptics – Quantitative suspension test for the evaluation of bactericidal activity of chemical disinfectants and antiseptics used in food, industrial, domestic, and institutional areas – Test method and requirements (phase 2 step 1)	Published.
BS EN 1650:1998	Chemical disinfectants and antiseptics – Quantitative suspension test for the evaluation of fungicidal activity of chemical disinfectants and antiseptics used in food, industrial, domestic and institutional areas – Test method and requirements (phase 2 step 1)	Published.
BS EN 13704:2002	Chemical disinfectants and antiseptics – Quantitative suspension test for the evaluation of sporicidal activity of chemical disinfectants and antiseptics used in food, industrial, domestic and institutional areas – Test method and requirements (phase 2 step 1)	Published.
BS EN 13697:2001	Chemical disinfectants and antiseptics – Quantitative non-porous surface test for the e valuation of bactericidal and/or fungicidal activity of chemical disinfectants and antiseptics used in food, industrial, domestic and institutional areas – Test method and requirements without mechanical action (phase 2 step 2)	Published.
prEN 14583	Workplace atmospheres – Volumetric bioaerosol sampling devices – Requirements and test methods	Under approval.
I/IEC 61010-1:2001	Safety requirements for electrical equipment for measurement, control, and laboratory use Part 1: General requirements	Published.

12

Definition of terms

A list of preferred terms used in isolator technology, associated devices and systems.

Scope

This guide uses many terms which are understood by the contributors and editors but are not necessarily interpreted in the same way by all readers. Individual experience and background may bring an interpretation which could be different. Therefore to avoid misunderstanding, the following definitions apply throughout the text of this book. Where possible, definitions are adapted from those in published standards, in which case the references are given. Strict definitions may still leave the reader confused about their meaning so some additional notes are provided as further guidance for some definitions.

Access device: A device for manipulation of processes, tools or products within the isolator (Draft ISO 14644-7).

This is the means whereby the operator can interact with processes in the controlled workspace. Gloves, gauntlets and robots are types of access device.

Action level: Microbiological or physical levels set by the user in the context of controlled environments. When exceeded, immediate follow-up is required as well as investigation with subsequent corrective action (Draft ISO 14644-7).

When action levels are exceeded, the isolator is running outside specification.

Aeration: The process of flushing the isolator controlled workspace with sterile filtered air until the concentration of a gaseous sanitising agent reaches a safe level.

This is the final part of the sporicidal gassing cycle and essentially eliminates the gassing agent from the controlled zones.

Alarm: An audible and/or visible and/or remote signalling system which warns of a fault condition.

Alarm (latched): See *Latched alarm.*

Alert level: Microbiological or physical levels set by the user for controlled environments giving early warning of a potential drift from normal conditions: when exceeded an investigation is required to ensure that the process and the environment are under control (Draft ISO 14644-7).

When alert levels are exceeded, action is required to ensure that action levels are not reached and normal levels are restored.

Anthropometrics: Study of the human anatomical dimensions with respect to the operation or management of a process.

Aseptic environment: An environment which is regulated to control microbial and particulate contamination to acceptable levels (Draft BS EN 13824).

This is an environment in which sterile starting materials and components are manipulated to produce sterile medicinal products with minimal risk of contamination.

Aseptic filling: Part of aseptic processing where a pre-sterilised product is filled and/or packaged into sterile containers and closed (Draft BS EN 13824).

Authorised Pharmacist: The person designated in writing by the Responsible Pharmacist to supervise the aseptic process and to release the product for use.

Background environment: The environment in which the isolator is sited.

More information on appropriate background environments is provided in Chapter 5.

Bioburden: Presence of viable microorganisms on a surface or within a space.

The number and type of viable microorganisms on a surface, in a medical device or package or pharmaceutical product or within a defined space (Draft BS EN 13824).

Biocide: A biocide is a general term (not actually used in this book) describing a chemical agent (in a liquid or gaseous form) that inactivates microorganisms.

This term is often applied to cleaning agents but should only be used for cleaning agents which have a biocidal action.

Biodecontamination: A general term used to cover the removal of bioburden by any process including physical removal, cleaning, disinfection and sanitisation.

Biological decontamination: A process of treating surfaces for removal, inactivation, or otherwise rendering harmless, biological contaminants.

Biological indicator (BI): A carrier inoculated with specific microorganisms that provide a defined resistance to a given antimicrobial process.

In isolators, BIs are normally used in the development of gassing cycles.

Blood labelling: The process of attaching to or tagging blood cells with an agent which can be used to identify such cells using an appropriate detection system.

Breach velocity: Air flow rate through an aperture sufficient to prevent movement of airborne particles in the opposite direction to the flow (Draft ISO 14644-7).

Breach velocity usually applies to a glove port where a breach may occur as a result of losing a glove. The accepted level is 0.7 m s^{-1}*, which is the value specified for Class III microbiological safety cabinets in BS EN 12469.*

Calibration: The set of operations which establish, under specified conditions, the relationship between values indicated by a measuring instrument or measuring system, or values represented by a material measure, and the corresponding known values of a reference standard (EC GMP).

Calibration is, in effect, the checking of fitted instruments and gauges and test instruments and gauges against reference values or standards to ensure that they are giving correct readings.

Change control: Formal procedure for recording **ALL** implemented alterations to documents, equipment, processes, procedures or specifications.

Chemical decontamination: The removal or rendering harmless of chemical contamination by any process including physical removal, cleaning, neutralisation or decomposition.

This is part of a decontamination process which reduces chemical contamination to a defined acceptance level. Examples of chemical contamination include spillages of cytotoxic drugs, excipients or active ingredients. In addition, filters used in the isolator may require decontamination before handling or disposal.

Chemical indicator: A device or kit that gives a qualitative indication of the presence or absence of a specified chemical.

In isolators, they are normally used in the development of gassing cycles. Chemical contamination monitors can be used to detect the presence of some chemical contaminants.

Chief Pharmacist: This is the pharmacist responsible for the pharmacy services within a corporate body.

CIP (Clean In Place): A remote controlled, semi-automatic or automatic, validatable, built-in system for cleaning surfaces.

The key requirement is that the CIP process can be validated.

Classified environment: An environment that is categorised according to levels of cleanliness, to a recognised classification.

Levels of operational cleanliness are specified in EC GMP. Levels of airborne particulate cleanliness are specified in BS EN ISO 14644-1.

Clean area: An area with defined environmental control of particulate and microbial contamination constructed and used in such a way as to reduce the introduction, generation and retention of contaminants within the area (EC GMP).

This will normally be in compliance with GMP grades for clean area.

Cleaning: Removal of contamination (which may include dust, soil, organic or inorganic matter) by physical means or by suitable agents from a surface to render it visibly clean.

The presence of surface contamination may inhibit the effectiveness of subsequent microbiological decontamination and disinfection processes.

Cleanroom: Room in which the concentration of airborne particles is controlled, and which is constructed and used in a manner to minimise the introduction, generation, and retention of particles inside the room, and in which other relevant parameters, e.g. temperature, humidity, and pressure, are controlled as necessary (BS EN ISO 14644-1).

Clean-up time: The time taken for a zone or area to recover from the operational or 'dirty' state to a specified at rest state or 'clean' state.

The term is normally applied to the clean-up time in a cleanroom after operations have ceased. The EC GMP expects a clean-up time of 15–20 min.

The same term is used for the isolator controlled workspace, in which case the clean-up time depends on the isolator design and its application.

The term can also apply to air-purged isolator transfer devices. These prevent the transfer of airborne contamination during transfers and the clean-up time should be very short to allow for quick, effective transfers. The time required for disinfectant action and evaporation is another factor in air-purged transfer devices.

Closed procedure: A procedure whereby a sterile pharmaceutical product is prepared by transferring sterile ingredients or solutions to a pre-sterilised sealed container, either directly or using a sterile transfer device, without exposing them to the external environment.

Compounding: A process wherein bulk drug substance is combined with another bulk drug substance and/or one or more excipients to produce a drug product.

Contaminating substances; Contaminants; Contamination: Materials which may be inanimate or of biological origin which are not necessary or part of the process.

Any such substances found in the controlled workspace should not exceed an acceptable level.

Containment: The provision of operator protection by means of a physical or aerodynamic barrier.

Controlled workspace: The enclosed workspace of an isolator constructed and operated in such a manner and equipped with appropriate air handling and filtration systems, to reduce to a predefined level the introduction, generation and retention of contaminants within it.

Cross-contamination: Contamination from other products or processes at the same process time or during earlier operations. (See also Process-generated contamination.)

Critical zone: That part of the *controlled workspace* of an isolator where containers can be opened and product may be exposed.

Decontamination: A process which reduces *contaminating substances* to a defined acceptance level.

Decontamination can encompass the following processes: biological, chemical and physical decontamination, cleaning, disinfection and sanitisation. Cross-contamination from earlier operations is dealt with by appropriate decontamination.

Design qualification (DQ): The documented verification that the proposed design complies with the URS and is suitable for the intended purpose.

This may involve a check of the manufacturer's calculations and supporting documents. It verifies and documents that the design, including the functional design specification (FDS), complies with the user requirement specification (URS).

Detergent: A cleaning agent that has wetting and emulsifying properties.

It is used to aid the removal of residues and soiling from a surface, leaving it visibly clean.

Disinfection: The process of reduction of the number of viable microorganisms in or on an inanimate matrix, by the action of an agent on their structure or metabolism, to a level judged to be appropriate for a specified, defined purpose.

The term is most commonly used to refer to liquid sanitisation.

Docking device: A transfer device that takes the form of a sealable chamber which can be completely removed from or locked onto an isolator and then opened without *contamination* passing into or out of the *controlled workspace* or the chamber.

DOP: Dispersed oil particulate (originally dioctylphthalate).

This is used in an aerosol generator to provide a challenge for leak testing of filters. The oil used to generate the challenge is usually a light mineral oil recommended by a specialist in filter testing.

D value: A term used in microbiology to indicate the decimal reduction time. This is the time required at any given temperature to reduce the viable count by 1 log order or 90%.

Ergonomics: The study of man in relation to his working environment.

This is frequently related to the adaptation of machines and general conditions to fit the individual so that he may work at maximum efficiency.

Exhaust filter: A filter through which the exit stream of air from an isolator passes.

Failure mode and effects analysis (FMEA): A method of assessing the risk posed by an action, process or event.

A scoring method is used to analyse the risks (see Chapter 10).

FDA: Food and Drugs Administration. The American regulatory authority controlling the quality of drugs, their manufacturing standards and other related issues. Equivalent to the UK MHRA (formerly MCA) (q.v.).

Fumigation: The delivery of a disinfectant to a device such as the internal controlled workspace of an isolator by aerial dispersion, usually in the form of a gas or vapour.

This is a term that is commonly used as an alternative to gassing when formaldehyde is the sanitisation agent.

GAMP: A guide prepared by ISPE (International Society of Pharmaceutical Engineers) to provide guidance that 'aims to achieve validated and compliant automated systems meeting all current healthcare regulatory expectations, by building upon industry good practice in an efficient and effective manner' (GAMP).

Although the title suggests that GAMP applies to automated systems (which includes control systems and software), it is written in such a way that it is a very useful guide to validation generally.

Gassing: A sporicidal process achieved by the distribution of a gaseous sanitisation agent to a defined space by dispersion in the form of a gas or vapour (see also fumigation).

HEPA: High efficiency particulate air filter.

HEPA filters are used to remove particulates from the supply air and exhaust air of isolators.

HVAC: Heating, ventilation and air conditioning.

An abbreviation very commonly used by design specialists and engineers.

Isolator: A separative device as defined in Draft ISO 14644-7 and used for pharmaceutical and related applications. It utilises constructional and/or aerodynamic means to enclose a *controlled workspace.*

*An isolator is **not** a sterilising device.*

Installation qualification (IQ): This should establish and document that the equipment has been correctly supplied and installed.

Laminar flow: Airflow in which the entire body of air within a defined zone moves with uniform velocity along imaginary parallel flow lines.

It should be noted that although this term has been in common usage for clean air systems, it has a different meaning in other areas of science and engineering such as aeronautical engineering and fluid dynamics. It has therefore been superseded by the term 'unidirectional flow'.

Latched alarm: An alarm system which continues to indicate an alarm condition until it is acknowledged and reset by an operator even if an alarm condition has passed.

MCA: Medicines Control Agency, which was merged in 2003 with MDA (Medical Devices Agency) to become the MHRA (q.v.).

Medical devices: A range of materials or products used in association with the medical treatment of patients and controlled and classified by the MHRA (q.v.).

MHRA: Medicines and Healthcare products Regulatory Agency. The MHRA was formed from a merger of the Medicines Control Agency (MCA) and the Medical Devices Agency (MDA) on 1 April 2003.

Most penetrating particle size (MPPS): The size of particle which is most likely to penetrate a HEPA filter.

HEPA filters retain larger and smaller particles by their various mechanisms. The size most likely to penetrate the filter is known as the MPPS. Filter manufacturers report the penetration efficiency at this size.

Negative pressure isolator: An *isolator* operating at a pressure less than atmospheric with the objective of protecting the product from process generated and external factors that would compromise its quality, and with the additional objective of protecting the operator from hazards associated with the product during operation and in the event of a failure of the isolator.

Non-unidirectional flow: Air distribution where the supply air entering the clean zone mixes with the internal air (BS EN ISO 14644-4:2001).

This is the same as turbulent flow.

Operator protection: Protection of the operator and the surrounding environment from hazards arising from the process or activity.

This is normally achieved by some form of containment.

Operational qualification (OQ): The OQ must show that the equipment can operate according to the URS and manufacturer's specification.

Parenteral administration: The administration of a medicine by a route other than via the alimentary tract. In practice this tends to mean administration of drugs by a direct route into the circulatory system.

Parenteral products need to be sterile as they bypass the body's natural barriers.

Performance qualification (PQ): This should demonstrate and document that the properly designed, installed and operating isolator can, with reliability and consistency, be used for the purpose for which it was designed.

PLC (Programmable logic controller): A type of small, self-contained computer which can be programmed to monitor and control many parameters.

Positive pressure isolator: An *isolator* operating at a pressure greater than atmospheric with the objective of protecting the product from process generated and external factors that would compromise its quality. A degree of operator protection is also achieved as long as a physical barrier is maintained.

Process-generated contamination: Contamination from other products or processes at the same process time or during earlier operations. (See also Cross-contamination.)

Product protection: The protection of products from viable and non-viable contamination arising from the operator and/or the surrounding environment.

Qualification: The process to demonstrate the ability to fulfil specified requirements (GAMP).

Radiopharmaceuticals: The group of pharmaceutical preparations which are specialised to incorporate a radionuclide for the purpose of patient treatment or diagnosis.

These substances emit ionising radiation and require special consideration to ensure safe handling whilst reducing the risk of product contamination.

Risk assessment: A process which takes account of all the related factors of a product and associated processes and evaluates the likelihood of failure of products to comply with specifications or operators to encounter higher than acceptable levels of exposure to hazards such as hazardous substances (see FMEA).

RTP (rapid transfer port): A transfer device in the form of a double-door transfer port system used to move items from one isolator to another without contamination entering or escaping from the system. The acronym is not normally applied to split-butterfly valves.

Safe change facility: A system which enables filters or filter systems to be changed without hazard to the operator and the background environment.

Sanitisation: The process of reduction in the number of micro-organisms in a space or in or on an inanimate matrix, by the action of an agent on their structure or metabolism, to a level judged to be appropriate for a specified, defined purpose.

It is the reduction of the level of viable microorganisms to a low level, or the killing of microorganisms that pose a threat to public health, or to a defined level, or by a defined log reduction. It does not necessarily mean the death of all microorganisms.

The term can be used as a semi-quantitative reduction of viable organisms by a factor of 10^6. It is usually achieved by surface decontamination using alcoholic sprays, swabs or dunking, as well as gaseous fumigation of controlled zones within an isolator system. It is normal to validate the process to determine its effectiveness.

Separative device: Equipment using constructional and dynamic means to create assured levels of separation between the inside and outside of a defined volume (Some industry specific examples of separative devices are clean air hoods, containment enclosures, glove boxes, isolators and minienvironments.) (Draft ISO 14644-7.)

Pharmaceutical isolators come within the scope of this generic standard.

Single hole equivalent (SHE): The size of a theoretical single hole which would account for all of the leakage observed during a leak test.

SIP (Sterilise in Place): A *sterilisation* system that sterilises equipment or an enclosure without first dismantling the equipment or enclosure.

Sporicidal gassing: The reduction of bacterial spores to an effective zero level using a gaseous agent.

Standard operating procedures (SOP): A document setting out the specific steps to be taken for a particular activity.

It is used to ensure consistency and facilitate training and it may also be referred to as a work instruction.

Sterile: Free of any viable organisms. (In practice, no such absolute statement regarding the absence of microorganisms can be proven, see *sterilisation*.)

Sterilisation: (BS EN ISO 14937 and ISO 11139).

The reduction of viable microorganisms to an effective zero level *or* validated process used to render a product free of viable organisms (PICS).

The complete destruction or removal of viable microorganisms, including bacterial spores, normally achieved by controlled application of heat, suitable electromagnetic irradiation, filtration or ethylene dioxide.

Note: *None of these processes are currently applicable to isolators.*

In a sterilisation process, the nature of microbiological inactivation or reduction in numbers of viable organisms is described by an exponential function. Therefore, the number of microorganisms which survive a sterilisation process can be expressed in terms of probability. While the probability may be reduced to a very low number, it can never be reduced to zero but can be expressed as a sterility assurance level (SAL). A SAL is the probability of a single viable microorganism occurring on a surface after sterilisation and is normally expressed as 10^{-n}.

An isolator is not generally accepted as a sterilisation device. Sterilisation can be achieved by physical and/or chemical means. It is currently not necessary to ensure that isolators are 'sterilised'. This would include validation to demonstrate a SAL in each zone of the isolator.

Sterility: The absence of living organisms. The conditions of the sterility test are given in the European Pharmacopoeia (PICS).

Sterility assurance level (SAL): Probability that a batch of product is sterile *or* the degree of assurance with which the process in question renders a population of items sterile (PICS).

It is expressed as the probability of a non-sterile item in that population. It is established by appropriate validation studies. (SAL is expressed as 10^{-n}*.)*

Thimble exhaust: An arrangement of an exhaust air system that draws air from both the isolator and the room in such a way that, when the isolator is switched on, the exhaust air is taken from the isolator preferentially.

This method is commonly used in negative pressure microbiological containment rooms but is equally applicable to isolators.

Transfer chamber: A particular type of transfer device which facilitates the transfer of goods into or out of the *controlled workspace* whilst minimising the transfer of contaminants.

Transfer device: A device which can be fixed or removable and which allows materials to be transferred into or out of the *controlled workspace.*

Transfer devices are usually designed to control contamination levels to a minimum and are classified into a number of different types, A1, A2, B1, B2, C1, C2, D, E, F, details of which are provided in Chapter 3.

Transfer hatch: This is a commonly used term that should not be used in isolator technology.

Transfer isolator: A separate isolator which can be fixed or removable and which is attached to the main operational unit, acting as a complete transfer device.

Turbulent flow: Airflow in which a body of air within a defined zone moves in a random directions.

See also Non-unidirectional flow.

ULPA filter: Ultra low particulate air filter.

These filters comply with the grades specified in the table given in Chapter 2.

Unidirectional flow: Controlled airflow through the entire cross section of a clean zone with a steady velocity and approximately parallel streamlines. Note: This type of airflow results in a directed transport of particles from the clean zone (BS EN ISO 14644-4:2001).

The term unidirectional flow has superseded the term laminar flow.

User requirement specification (URS): The key defining document against which all qualification to verify compliance is based.

The URS is a requirement specification that describes what the equipment or system is supposed to do, thus containing at least a set of criteria or conditions that have to be met (GAMP).

It is a specification devised by the user, with or without assistance from a specialist, that defines what is required by the organisation which is to use the isolator.

Validation: The accumulation of documentary evidence to show that a system, equipment or process will consistently perform as expected to a predetermined specification, and will continue to do so throughout its life cycle.

It establishes documented evidence which provides a high degree of assurance that a specific process will consistently produce a product meeting its predetermined specifications and quality attributes (GAMP/FDA).

Validation master plan (VMP): A coordinating document describing the validation of a total system comprising individual pieces of equipment and/or processes.

The VMP should begin with policy and strategy for total system validation and show how different items of equipment and processes are to interact to form a total system. It should list all associated validation documents including individual validation plans and protocols, and should include those documents in existence and those to be created to complete the validation study.

Appendix 1

'Handling cytotoxic drugs in isolators in NHS pharmacies'

HSE/MCA January 2003

This appendix is a copy of the guidance for good practice produced by a joint working party of the NHS, MHRA (formerly the MCA) and HSE to help identify the key points in the manipulation of cytotoxic drugs in positive and negative isolators. The original, which is reproduced here, is available from the HSE or MHRA or searching for by the title in full on an internet search engine.

A1.1 Introduction

1 This joint Health and Safety Executive (HSE)/Medicines Controls Agency (MCA) guidance gives advice on factors to consider when selecting either negative or positive pressure isolators for the aseptic reconstitution of cytotoxic drugs. The guidance is aimed at:

- pharmacy managers;
- quality control managers;
- those responsible for training staff;
- health and safety advisers;
- employee safety representatives; and
- those responsible for supplies and purchasing.

The isolator has to perform two functions. It is a key control measure in preventing employee exposure to cytotoxic drugs, many of which are classified as hazardous to health and may also be carcinogens. It also has to protect the product from microbiological contamination during drug reconstitution. This guide will help those responsible for selecting islolators to choose the type of isolator appropriate for both these purposes. It is not intended to give guidance on other aspects of safe systems of work in the pharmacy.

2 Both positive and negative pressure isolators are enclosed systems and rely on a steady flow of filtered air during use. A slight pressure differential is placed on the isolator, either negative or positive. Isolators are intended to eliminate or control the operator's exposure to the cytotoxic drug during reconstitution, as required by the Control of Substances Hazardous to Health Regulations 1999.[1] In addition, isolators reduce the potential for microbial contamination of the product, as specified in the *European Guide to Good Manufacturing Practice*.[2]

3 Negative pressure isolators are designed to give optimal protection to the operator. Positive pressure isolators are designed to enhance product protection. Air entering and leaving the isolator, whether positive or negative, will do so through the HEPA filters. A leak on the isolator, such as hole in the isolator wall or a defective seal, will allow air to bypass the HEPA filters and to directly leave or enter the system. For a positive pressure system, this will allow air that may be contaminated with cytotoxic drug to enter the workplace. For a negative pressure system, air that may contain bacteria could enter the isolator and contaminate the preparation. If the breach is obvious, the isolator should be taken out of use until it is repaired. A good leak detection regime will ensure that the presence of such defects, whether obvious or not, are identified as early as possible.

4 The period of time between loss of integrity of the system and detection of the leak is crucial. It is extremely important that the early detection and repair of leaks is given particular attention. But, this is not the only source of operator exposure or of product contamination. It is important to assess all potential sources of operator exposure and contamination and take appropriate steps to minimise risks to worker and patient health.

5 A recent HSE study in two pharmacy units, one using positive pressure and one using negative pressure isolators,[3] found no significant difference in operator exposure to cytotoxic drugs between the units. This exposure was measured as surface contamination and airborne concentrations. Evidence of absorption by operators was studied by analysis of drugs or their metabolites. These exposures and the measured absorption were significantly lower than previous published studies, suggesting that a correctly designed, validated and maintained isolator can reduce the risk to the operator, irrespective of the pressure differential.

6 This was only a limited study, but it would seem that in well-managed units, the low levels of exposure and absorption measured were a consequence of factors other than the pressure of the isolators. Only with a significant fault, would the pressure of the isolator have a major impact on the operator exposure.

A1.2 Routes of operator exposure

7 Operators can be exposed to cytotoxic drugs through factors such as:

- breathing air contaminated with cytotoxic drug as a powder, or aerosol or vapour;
- skin contact with the drug itself or contaminated surfaces, some of these drugs can pass through intact skin;
- accidental ingestion.

Isolator selection to achieve control of worker exposure and product protection has to be a local decision based on factors such as those in Appendix 1. It is the full package of control measures that will achieve a high standard of control with either type of isolator. An essential prerequisite for adequate control of both exposure and contamination is a well-trained workforce who are skilled in how to deal with both routine manipulations and the action to take if there is a major leak or spillage inside the isolator. This training needs to be conducted on a regular basis and to be updated when any major change is made, to reflect major changes to procedures and to ensure that competence levels are maintained.

8 HSE and MCA cannot stipulate which type of isolator to select. It is possible to use either positive or negative pressure isolators to maximise drug protection and minimise employee exposure. Factors affecting worker health and drug protection should be fully taken into account by means of documented risk assessment, failure modes and human error analysis, together with rigorous change control. Pharmacy workers and their representatives should be involved in these processes.

9 This publication is intended to help in this selection procedure, and to give advice on safe use, for both types. The final choice of which type of isolator to use, is dependent on a range of factors. These are discussed in paragraphs 10–19.

A1.2.1 Factors involved in employee exposure or product contamination

A1.2.1.1 Factors common to both employee exposure and product contamination

Routine maintenance procedures for the isolator such as glove changes, cleaning of the isolator and filter changes

10 Regular changes of the isolator gloves are essential and this must be performed in a way which minimises possible contamination. Safe systems of work (or safe operating procedures) should be established for changing exhaust HEPA filters.

A significant leak through the containment layers of the isolator

11 This is where the pressure of the isolator may have a considerable impact. Loss of integrity in a negative pressure isolator, ie an inward flow of air, is less likely to give rise to operator exposure, but may cause microbiological contamination of the product. In positive pressure isolators, although some protection is provided, a leak may overwhelm the effect of the positive pressure and compromise the product.

12 A significant leak from a positive pressure isolator may lead to contamination of the operator and the immediate environment. Alarm systems for positive pressure isolators need to be sensitive and allow isolator shutdown and rapid evacuation of the room, before any significant exposure occurs. Investigation of the cause of the alarm should be investigated by people wearing personal protective equipment, which is both suitable and sufficient. Gloves need to resist both permeation and penetration of the drug. Only operators fully trained in the use of this equipment should participate.

13 A significant leak from a negative pressure isolator still requires evacuation of the room.

14 A safe operating procedure for dealing with alarms should be established, including decontamination procedures. Procedures should be practiced at regular intervals.

Natural leakage through the isolator

15 This is particularly important for positive pressure systems where any such leakage may result in the escape of cytotoxic drug from the isolator. The significance of the leak will depend on the amount of air escaping and the concentration of the cytotoxic drug in the

air. Therefore, you should ensure working practices minimise release of cytotoxic drug into the isolator atmosphere and the isolator is adequately maintained to minimise leakage. Your leak detection system should then be able to detect low level losses from the isolator.

16 The higher the rate of airflow through the isolator, at constant pressure differential, the lower the residence time of air inside, and the steady state concentration of drug is reduced. A minimum of 40 air changes per hour is normally required, but different designs may enable adequate ventilation at lower air change rates.

A1.2.2 Factors specific to employee exposure

Reconstitution of the drug

17 Transfer of fluid to a vial containing drug may overpressurise the vial, resulting in the release of air containing cytotoxic material into the inside of the isolator. The actions necessary to remove air bubbles from a syringe may also result in release of contamination. The contaminants can be in the form of an aerosol or vapour. These activities will be the major source of release of cytotoxic drug into the isolator atmosphere. Therefore every effort should be made to adopt techniques and working practices that minimise releases during reconstitution (and any other transfer activities). Achieving this will reduce the significance of emissions of cytotoxic drug into the local atmosphere, which may occur if there is a leak from the isolator. Such a release should be largely removed by the extraction of the isolator. If it remains inside the isolator, it may deposit on internal surfaces or on transient materials passing through the isolator.

Contaminated surfaces

18 Some cytotoxic drugs can pass through intact skin and this could be a major route of entry into the body. Failure to wear adequate personal protective equipment, such as clean and undamaged gloves inside the isolator gloves, may expose individuals to cytotoxic drugs. Control any activities that result in the release of cytotoxic drug into the isolator or pharmacy carefully to minimise these. Instigate a cleaning regime of appropriate frequency and standard that prevents contamination build-up. Consider periodic testing of work-place hygiene practices by undertaking surface wipe sampling.

A1.2.3 Factors specific to product contamination

19 The isolator provides an environment in which aseptic manipulations are carried out. If micro-organisms are present in this environment, they may contaminate the product when sterile surfaces and materials are exposed. Micro-organisms may be present or gain access by the following routes.

- Survival of the cleaning and sanitisation process applied to the resident surfaces in the isolator. If a sporicidal gassing process is used this is less of a risk.
- Transfer in on the surfaces of transient materials passing through the isolator. This can occur even though surfaces are sanitized. If a sporicidal gassing process is used, this is less of a risk.
- Transfer into, and contamination of the isolator environment by using non-sterile materials. These may include non-sterile raw materials, non-sterile equipment, non-sterile fluids, non-sterile vacuum connections, non-sterile gases, non-sterile lubricants for door seals etc. Note that if non-sterile materials are components of the product, the product will be non-sterile, but this is not a specific consequence of using an isolator.
- Ingress through the physical barriers that comprise the isolator. These include:
  - Failure of inlet and outlet HEPA filters.
  - Loss of integrity of the operator contact parts of the isolator, such as gloves, sleeves and suits. A positive pressure may not provide protection in these circumstances. Negative pressure may actively draw contaminants into the isolator. In the case of loss of integrity of non-operator contact parts, positive pressure provides some protection, whereas negative pressure will deliver any contaminants in the surrounding room air to the isolator.

A1.3 Combining risk to operator with risk to product

20 As stated previously, there is much more to consider than merely the pressure differential of the system. If the above sources of exposure and sources of product contamination can be minimised, then the type of system selected should be less important. This

assumes that there is no catastrophic leakage. In this case, alarm systems and training systems become paramount.

21 You should now examine the table in Appendix 1 and consider what best suits your pharmacy needs. The table describes the consequences for positive and negative pressure isolators on the critical performance factors for their use. Each type of isolator will bring in some extra specific precautions, and it is up to the pharmacy to make the decision in the knowledge of what it will entail. It is recognised that the type used is very much dependent on the exact needs of your pharmacy. Therefore the table describes good practice for both types, so that you can ensure safe standards are met when you have made your choice.

22 Operator protection advantages of negative pressure and the product protection advantages of positive pressure can be combined in one isolator; this 'double skin' technology is available.

23 If you need further advice after reading this document, the following sources are recommended:

- Regional QC pharmacist
- Medicines Control Agency general enquiry point:
  0207 273 0000 MCA specific isolator enquiry,
  Andrew Bill: 01904 610556
- Health and Safety Executive Infoline: 08701 545500

A1.4 Negative and positive pressure: Decision table

The main purpose of this table is to draw attention to the extra considerations arising from a decision to use either a positive or a negative pressure isolator. Where there is no comment, this does not mean that a feature is not important. Other important features may not be detailed below.

Appendix 1 Decision table

NEGATIVE PRESSURE	*FACTOR*	*POSITIVE PRESSURE*
Product protection – There may be specific requirements for the standard of the pharmacy air that may be drawn into the isolator. Usually Grade D is expected provided that leak detection is carried out as described in the 'Leak detection/testing' section on page 5. The room pressure should be the minimum required for Grade D. *Operator protection* – There are no additional air quality standards for pharmacy units above those required for any workplace.	*Pharmacy environment*	*Product protection* – Leaks will tend to result in air escaping the isolator, therefore the standard of the pharmacy air become less important. Usually Grade D is expected. *Operator protection* – There are no additional air quality standards for pharmacy units above those required for any workplace.
Product protection – Hatches and other transfer devices must be designed to prevent unfiltered air from entering the working zone(s) both in use and at rest. *Operator protection* – Hatches and other transfer devices must be designed to prevent potentially contaminated air from leaving the working zone(s) and entering the room in which the operators are working.	*Transfer devices*	*Product protection* – Hatches and other transfer devices must be designed to prevent unfiltered air from entering the working zone(s) both in use and at rest. *Operator protection* – Hatches and other transfer devices must be designed to prevent potentially contaminated air from leaving the working zone(s) and entering the room in which the operators are working. With positive pressure isolators, there is much more potential for this to happen. If positive pressure is used, the standard of transfer devices needs to be higher to ensure that this does not occur.

Appendix 1 Decision table *continued*

NEGATIVE PRESSURE	FACTOR	POSITIVE PRESSURE
Product protection – The use of aseptic techniques, correctly devised with regard to the direction of laminar airflow, is expected to provide a reduction in the risk that any micro-organisms that may be present would contaminate the product. Turbulent airflow does not provide this element of reduction in risk. It should be noted that neither laminar nor turbulent airflow should be assumed to deflect the high velocity jet of potentially contaminated air entering the isolator through a leak. *Operator protection* – Whether laminar or turbulent airflow is used the air should effectively scour the space inside the isolator and remove any airborne drug that may be released during operations.	*Laminar or turbulent airflow*	*Product protection* – The use of aseptic techniques, correctly devised with regard to the direction of laminar airflow, is expected to provide a reduction in the risk that any micro-organisms that may be present would contaminate the product. Turbulent airflow does not provide this element of reduction in risk. *Operator protection* – Whether laminar or turbulent airflow is used the air should effectively scour the space inside the isolator and remove any airborne drug that may be released during operations.
Product protection – Minimum necessary to achieve containment objectives. *Operator protection* – The pressure differential should be sufficient to ensure the effective operation of the isolator during all foreseeable operating conditions including cleaning and maintenance, and sufficient to ensure that normal operating conditions do not overwhelm it. Negative pressure should be sufficient to generate a breach velocity of at least 0.7 m/sec.	*Pressure differentials*	*Product protection* – Sufficient to prevent pressure reversals and maintain at least 15 Pa at all times. *Operator protection* – The positive pressure differential should be as low as possible, but in line with product protection requirements.

Appendix 1 Decision table *continued*

NEGATIVE PRESSURE	*FACTOR*	*POSITIVE PRESSURE*
Product protection – Rigorous aseptic technique should be used on the assumption that micro-organisms may be present. *Operator protection* – Systems of work should minimize the generation of aerosols during drug reconstitution, and prevent drug contamination on the surfaces of vials and interior walls. This is irrespective of isolator type. Methods that minimise product transfer should be sought. Products requiring little or less manipulation should be considered.	*Systems of work*	*Product protection* – Rigorous aseptic technique should be used on the assumption that micro-organisms may be present. *Operator protection* – Systems of work should minimize the generation of aerosols as with negative pressure systems. However this becomes more important as any leaks may result in contaminated air escaping from the isolator. Methods that minimise product transfer should be sought. Products requiring little or less manipulation should be considered.
Product protection – Training in the special risks regarding leaks. *Operator protection* – Operators should receive adequate training in the hazards and risks of the materials they work with and the steps needed to minimise those risks. This should include the actions to take if a leak is found, evacuation drills and decontamination procedures.	*Training programmes*	*Product protection* – Standard GMP and guidance on isolators. *Operator protection* – Operators should receive adequate training in the hazards and risks of the materials they work with and the steps needed to minimise those risks. This should include the actions to take if a leak is found, evacuation drills and decontamination procedures.
Product protection – During installation/qualification carry out distribution leak test including arms and gloves. Limit: individual leaks 20 micron. The pressure decay limit determined in this state sets the limit for routine use. See note 1 on page 201.	*Leak detection/ testing*	*Product protection* – During installation/ qualification carry out distribution leak test including arms and gloves. Limit: individual leaks 20 micron. The pressure decay limit determined in this state sets the limit for routine use. See note 1 on page 201.

Appendix 1 Decision table *continued*

NEGATIVE PRESSURE	*FACTOR*	*POSITIVE PRESSURE*
Product protection – Identification and monitoring (particulate and micro) of possible inleak sites. More intensive control and monitoring of the surrounding room.	*Monitoring systems*	*Product protection* – Monitoring as appropriate for isolators.
Product and operator protection – Gated alarms as necessary.	*Alarm systems*	*Product and operator protection* – Gated alarms as necessary.
Product and operator protection – The COSHH Regulations require that isolators are properly maintained and undergo a thorough examination and test at least once every 14 months. This periodic check should be complemented by regular checks of the system. This may include daily visual checks of the condition of the isolator (in particular any obvious holes or other defects) and pressure gauge readings. These measures would be in addition to routine leak testing.	*Other maintenance procedures*	*Product and operator protection* – The COSHH Regulations require that isolators are properly maintained and undergo a thorough examination and test at least once every 14 months. This periodic check should be complemented by regular checks of the system. This may include daily visual checks of the condition of the isolator (in particular any obvious holes or other defects) and pressure gauge readings. These measures would be in addition to routine leak testing.
Product protection – Pinhole breaches in gloves can present the opportunity for air to enter the isolator at sufficient velocity to compromise product. Visual inspection for leaks before starting operations and systematic examination throughout the day is necessary. It is important that only well-fitting gloves are used to avoid ballooning.	*Routine use of isolator gloves*	*Product protection* – Pinholes in the gloves are a potential problem irrespective of positive pressure. It is unlikely that positive pressure will transfer to a glove. Periodic systematic visual inspection is necessary. Accurate glove sizing is less critical.

Appendix 1 Decision table *continued*

NEGATIVE PRESSURE	*FACTOR*	*POSITIVE PRESSURE*
Operator protection – Holes in gloves still present a risk to the worker, although less than with positive pressure.	*Routine use of isolator gloves*	*Operator protection* – Permeation and penetration both need to be considered. Permeation (transport through the glove material) is unaffected by the air pressure. Penetration (leakage of drug through holes or through bad seals) will be increased by positive pressure. In these systems, examination of the glove integrity should be routinely carried out before the isolator is used.
Product and operator protection – A system must be in place that ensures that gloves are replaced at appropriate intervals. A safe system of work should be established to ensure that contamination of the worker does not occur during this operation.	*Isolator glove changing*	*Product and operator protection* – As for negative systems a system must be in place to ensure that contamination is prevented.
Product protection – Sanitised and impervious inner sleeves and clean inner gloves. Possible higher grade clothing. *Operator protection* – Clean gloves should be worn at all times and changed regularly at least every four hours.	*Operator clothing*	*Product protection* – Standard Grade D clothing. *Operator protection* – Clean gloves should be worn at all times and changed regularly at least every four hours.

Appendix 1 Decision table *continued*

NEGATIVE PRESSURE	*FACTOR*	*POSITIVE PRESSURE*
Product and operator protection – For in-house quality control purposes, it is possible to measure levels of some cytotoxic drugs in air[3, 4] or on surfaces. Biological monitoring involving, for instance, urine samples, is an option for quality control purposes also. However, these procedures need to be optional, involve consultation with employees and be subject to informed consent. *Biological monitoring in the workplace. A guide to its practical application to chemical exposure*[5] is available by mail order from HSE Books (see Further information for details).	*Additional procedures. Monitoring and surveillance*	*Product and operator protection* – For in-house quality control purposes, it is possible to measure levels of some cytotoxic drugs in air[3, 4] or on surfaces. Biological monitoring involving, for instance, urine samples, is an option for quality control purposes also. However, these procedures need to be optional, involve consultation with employees and be subject to informed consent. *Biological monitoring in the workplace. A guide to its practical application to chemical exposure*[5] is available by mail order from HSE Books (see Further information for details).

Note 1 During the installation qualification, a leak test with tracer gas or aerosol and detector will enable the leaks distributed in the isolator to be detected. The test should be sensitive enough to detect individual leaks of less than 20 microns. Once all leaks detected have been eliminated, the isolator can be subjected to the pressure decay test that is to be used routinely. The pressure decay found in this test sets the limit for the routine test. The pressure decay test should include sleeves and gloves. Initially the test should be carried out daily until the stability of the integrity of the isolator is established. Following this, the frequency can be reduced to weekly.

References

1. *Control of substances hazardous to health. Control of Substances Hazardous to Health Regulations 2002. Approved Codes of Practice and guidance* L5 (Fourth edition) HSE Books 2002 ISBN 0 7176 2534 6
2. *The rules governing medicinal products in the European Community. Good manufacturing practices for medicinal products* European Communities/Union 1992 ISBN 9 28263180 X

3. Mason H *Cytotoxic drug exposure in two pharmacies using positive or negative pressurised enclosures for the formulation of cytotoxic drugs* Report No. HEF/01/01, HSL Sheffield
4. Ziegler E, Mason H, Baxter P 'Occupational exposure to cytotoxic drugs in two oncology wards' *J Occup Environ Med* 2002 **59** 608–612
5. *Biological monitoring in the workplace: A guide to its practical application to chemical exposure* HSG167 HSE Books 1997 ISBN 0 7176 1279 1

While every effort has been made to ensure the accuracy of the references listed in this publication, their future availability cannot be guaranteed.

Further information

HSE priced and free publications are available by mail order from HSE Books, PO Box 1999, Sudbury, Suffolk CO10 2WA Tel: 01787 881165 Fax: 01787 313995 Website: www.hsebooks.co.uk (HSE priced publications are also available from bookshops and free leaflets can be downloaded from HSE's website: www.hse.gov.uk.)

For information about health and safety ring HSE's Infoline Tel: 08701 545500 Fax: 02920 859260 e-mail: hseinformationservices@natbrit.com or write to HSE Information Services, Caerphilly Business Park, Caerphilly CF83 3GG.

This leaflet contains notes on good practice which are not compulsory but which you may find helpful in considering what you need to do.

Appendix 2

Training

A summary guide on training for operators who work with isolators. Information is presented as a checklist for adoption or adaptation in different workplaces.

A2.1 Introduction

Structured GMP training programmes should be provided for all staff working with isolators[1,2]. Training should be appropriate to grades of staff, and their responsibilities and duties.

Although training is given under general headings, it should be based on the SOPs that individuals are required to follow.

A2.2 Performance objectives

Training should meet the following performance objectives.

A2.2.1 Principles of basic theory, design, and siting of isolators

Operators should have an understanding of:

1. Principles of isolator design;
2. Different types of isolators;
3. Different types of transfer devices;
4. Airflow patterns inside isolators;
5. Background environments for isolators;
6. Entering, working in, and leaving the background environment.

A2.2.2 Procedures for the operation of isolators

Operators should understand, and be competent in following SOPs for:

1. Pre-use checks;
2. Entry, use and exit of access devices including gloves, sleeves, gauntlets or half-suit;
3. General operation;
4. Use of equipment inside isolators;
5. Maintenance and servicing;
6. Failure and troubleshooting.

A2.2.3 Procedures for materials transfer – non-gassed isolators

Operators should understand, and be competent in following SOPs for:

1. Surface sanitisation processes;
2. Transfer of materials into controlled and critical areas;
3. Transfer of materials out of controlled and critical areas;
4. Removal of waste including hazardous waste.

A2.2.4 Procedures for loading and sanitisation of gassed isolators

Operators should understand the principles of gaseous sanitisation, and the choice of agents used.

Operators should understand, and be competent in following SOPs for:

1. Loading patterns;
2. Displacement;
3. Appropriate temperature/humidity monitoring.

A2.2.5 Procedures for aseptic processing

Operators should understand, and be competent in following SOPs for:

1. Standard aseptic processing techniques;
2. Segregation and flow of products.

A2.2.6 Procedures for integrity testing

Operators should understand, and be competent in following SOPs for:

1. Inspection of gloves, sleeves, gauntlets and half-suits;
2. Leak testing of gloves, sleeves, gauntlets and half-suits;
3. Repairs of sleeves and half-suits;
4. Pressure decay tests;
5. Detection of leaks in all other parts of the system.

A2.2.7 Procedures for routine glove changing

Operators should understand the risks presented to the integrity of the system when carrying out these procedures.

Operators should understand, and be competent in following SOPs for:

1. Changing gloves;
2. Changing gauntlets;
3. Changing sleeves;
4. Changing half-suits.

A2.2.8 Procedures for decontamination

Operators should understand the need to validate cleaning and disinfection procedures with appropriate agents, and the potential need to vary agents used.

Operators should understand the difference between cleaning and sanitisation, and be competent in following SOPs for:

1. Cleaning and sanitisation of background environments;
2. Cleaning and sanitisation of isolators;
3. Decontamination of isolators after use of or spillage of hazardous materials.

A2.2.9 Environmental monitoring

Operators should understand potential sources of contamination. They should also understand what corrective actions they have to take in the event of alert levels and action levels being indicated.

Operators should understand, and be competent in following SOPs for:

1. Particulate monitoring;
2. Active microbiological monitoring;
3. Passive microbiological monitoring;
4. Monitoring of chemical contamination;
5. Radiation contamination monitoring.

A2.2.10 Safety

Operators should understand and be competent in SOPs for actions in the event of emergency, e.g. loss of a complete glove.

A2.2.11 Documentation and change control

Operators should understand all the documentation that they are required to complete. They should also understand the principles and procedures for change control. One of the principles of change control is that before any change is implemented, any staff affected by such change, receive appropriate training which should be documented as part of the change control process.

A2.3 Structure and assessment

A2.3.1 Under-pinning knowledge

1. Appropriate background reading;
2. Read SOP;
3. Appropriate practical demonstrations.

A2.3.2 Performance

1. Appropriate simulated work;
2. Appropriate practice under supervision.

A2.3.3 Assessment

The effectiveness of training should be assessed using any or all of the following methods:

1. Natural observation;
2. Observation of outcomes;
3. Process simulation;
4. Oral questioning;
5. Written questioning.

A2.4 Ongoing training and assessment

This should be undertaken after initial training, and at regular intervals.

After appropriate training and validation, individuals may be assessed as:

1. Operator – able to work to a SOP;
2. Trainer – able to train to a SOP;
3. Assessor – able to assess a trainee working to a SOP.

A2.5 Records

Training and assessment should be documented as it is done. Individual records should be maintained for each member of staff. This should include:

1. Performance objectives;
2. Understanding and working to SOPs (retraining is required if SOPs are modified);
3. Documentation of assessment and further training;
4. Date;
5. Name of trainer;
6. Trainee confirmation that training has been received.

A2.6 Other training

Staff may also require training in:

1. Health and Safety[3];
2. COSHH[4];
3. Handling of radioactive materials[5,6,7]; the preparation of radiopharmaceuticals requires that operators must have received 'adequate' training under The Ionising (Medical Exposure) Regulations[5].

A2.7 Practical note

It has been noted recently that process technicians or operatives are being asked to note data on batch records that might affect the product. Items which may need to be recorded include:

1. Airflow readings;
2. Pressure gauge readings;
3. Production line speeds;
4. Temperature and humidity readings.

The readings may be taken where there is no recording device fitted. In such circumstances, there is clearly a need for training in taking the readings properly, including the correct use of any measuring instruments. This is usually covered in training regimes, although there is sometimes a need for regular refresher courses.

There is also a need for training in the action to be taken when out-of-specification (OOS) results are observed. The importance of highlighting OOS readings and recording of any actions taken, should be emphasised.

References

1. *Rules and Guidance for Pharmaceutical Manufacturers and Distributors*, London, The Stationery Office Ltd, 2002.
2. Beaney, A. (ed.), *Quality Assurance of Aseptic Preparation Services* (3rd edn) London, Pharmaceutical Press, 2000.
3. *The Health and Safety at Work Act 1974*, London, HMSO, 1974.
4. SI 1999/No. 437 *Control of Substances Hazardous to Health Regulations 1999*, London, The Stationery Office Ltd, 1999.
5. SI 2000 No. 1059 *The Ionising Radiation (Medical Exposure) Regulations 2000*, London, The Stationery Office Ltd, 2000.
6. Allisy-Roberts, P. (ed.), *Medical and Dental Guidance Notes. A good practice guide to implementing ionising radiation legislation in the clinical environment*, York, Institute of Physics and Engineering in Medicine, 2001.
7. SI 1999 No. 3232 *Ionising Radiation Regulations 1999,* London, The Stationery Office Ltd, 1999.
8. MARC Panel, *Management and Awareness of the Risks of Cytotoxics* http://www.marcguidelines.com/

Appendix 3

Stainless steel for isolators

Stainless steel is widely used in isolators. Further understanding of this material might therefore be useful to those specifying, building and using isolators.

A3.1 Steel alloys

Stainless steels form a group of the iron alloy family called 'steel'. Stainless steels must contain at least 10.5% chromium. Within the family of stainless steels, there are three main metallurgical groups, martensitic, ferritic and austenitic. Pharmaceutical isolators should be constructed using stainless steel of the grades 316L or 304, which belong to the austenitic group. The two can quite easily be distinguished using specific reagents, referred to as the 'molybdenum test'. Proprietary kits are available for this test.

The austenitic stainless steels are characterised by their better corrosion resistance and their non-magnetic nature. The constitution of 316L and 304 grade stainless steels is given in Table A3.1.

Table A3.1 Stainless steel types and composition

Type	*316L*	*304*
Carbon	0.03% max	0.08% max
Chromium	16–18%	18–20%
Manganese	2% max	2% max
Molybdenum	2–3%	–
Nickel	10–14%	8–10.5%
Phosphorus	0.045% max	0.045% max
Silicon	1% max	1% max
Sulphur	0.03% max	0.03% max
Iron	balance	balance

The 18/8 or 18/10 grade stainless steels used in stainless steel cutlery contain 18% chromium and either 8% or 10% nickel, respectively, so could be classified as either 304 or 316L, respectively.

A3.2 Corrosion resistance of stainless steel

The chromium in the stainless steel has a great affinity for oxygen, and will form on the surface of the steel at a molecular level a film of chromium oxide. The film itself is about 0.13 μm in thickness. This layer is described as passive, tenacious and self-renewing. Passive means that it does not react or influence other materials; tenacious means that it clings to the layer of steel and is not transferred elsewhere; self renewing means that if damaged or forcibly removed more chromium from the steel will be exposed to the air and form more chromium oxide.

It is this passive layer of oxide that protects the steel from further corrosion and pitting; however, corrosion can still take place on stainless steels, under certain circumstances.

Even the best stainless steels will rust when exposed to solutions containing chloride ions; 316L has better resistance to attack than the 304 stainless steel. This tends to occur when the oxide layer is broken, e.g. by knocks and scratches, and the chloride ions are then able to penetrate and bind preferentially to the chromium, thereby breaking the integrity of the oxide layer. Once this occurs, the exposed iron atoms are vulnerable to the rusting process.

The process of rusting may be compounded by changes in the metallurgical structure of the stainless steel, which may have been caused by the welding process. 316L stainless steel is less prone to this 'sensitisation' effect than 316 stainless steel. The L in the 316L indicates that it is low carbon. Similarly, 304L may be less prone to this 'sensitisation' than 304 stainless steel. When sensitisation has occurred, corrosion can occur along the line of welding joint.

It is also worth noting that stainless steel may corrode after contact with mild steel materials or tools. The ferrous material transferred sets up a corrosion site on the stainless steel surface. The presence of ferrous contamination can be detected with a reagent test and, again, proprietary kits are available for this check.

316L stainless steel should be the isolator construction steel of choice if using chloride solutions at room temperature, since it confers much greater corrosion resistance under those conditions. Pharmaceutical isolators are often used to manipulate drugs in conjunction with normal saline solutions, so it would be wise to specify 316L stainless steel construction, at least for the critical areas that may come in contact with chloride. Examples are work surfaces and transfer devices.

To some extent, corroded stainless steel surfaces can be chemically cleaned or mechanically refinished, and then 'passivated' with proprietary solutions or pastes.

A3.3 Stainless steel finishing

The stainless steel used for the fabrication of isolators can be given a surface finish by a number of methods which impart significantly different properties to the surface. The surface finish can be specified and measured in a number of ways. It is important to understand the inter-relationship of the finishing methods and the specifications.

A3.4 Finishing methods

A3.4.1 Prefinished

Manufacturers will often use sheet steel that has been prefinished to a given standard. This can vary from simple passivation, through to mirror polish. Passivation is a process of treatment with acid to produce a tough oxide layer with a pearly grey finish.

A3.4.2 Hand finish and emery belt finish

These are both used to impart a grained finish which can vary from coarse, which is noticeably rough to the touch, through to almost mirror finish. The grain may all run in one direction resulting in a 'brushed' finish, or may be random, to give a slightly dull even finish. It is necessary to work from a coarse finish, through to a fine finish when using these processes.

A3.4.3 Electropolishing

This is a specialist electrochemical process, which actually removes 10–15 μm of material from the surface and leaves a very smooth, lustrous finish. It is to be recommended for pharmaceutical applications, especially where product contact is involved. The electropolished surface is easier to clean thoroughly than grained finishes and does not trap small particles in the surface. In particular, the method removes the so-called 'Beilby layer'. This is a very thin layer of debris left by all other finishing processes, and which can negate the properties of the underlying metal.

A3.5 Terminology

A3.5.1 Grit size

This is a measure of the abrasive particle size, used to apply the surface finish.

A3.5.2 Roughness average (Ra)

This is a measure of the actual 'ridges' and 'furrows' on the metal surface, measured in micrometres. The instrument used to determine Ra consists of a fine needle that is drawn across the surface while its movement is magnified and recorded. It is also known as 'centre line average' but the term Ra has now become widely used. This system is the preferred method for specifying pharmaceutical finishes. See Table A3.2.

A3.6 Application to pharmaceutical isolators

In general, the finest finish should be specified for isolators, but cost will always be a consideration. Most isolators currently have a finish in the region of Ra 0.8 µm. If requested, manufacturers should supply a series of actual Ra measurements made at various sites across the isolator carcass, and these should be retained in the IQ protocol.

A3.6.1 Very high corrosion resistant alloys

Where very high corrosion resistance is required, especially in acidic conditions, then a more specialised alloy may be specified. This is likely to come from a family of expensive alloys that are composed of nickel and chromium with additions of molybdenum, titanium, iron and

Table A3.2 Comparative table of finishes

Grit size (approx.)	*Roughness average (Ra; µm)*	*Typical description*
80	2.5	Coarse brush
120	0.8	Fine brush
180	0.4	Dull mirror polish
240	0.2	Mirror polish (electropolish)
320	0.1	Bright mirror polish
500	0.05	Optical quality

aluminium. To reduce the cost, the material can be welded to ordinary stainless steels to provide areas of specific resistance.

A3.6.2 Cleaning

316L stainless steel is less prone to corrosion than other stainless steels in the presence of chloride ions at room temperature. Where spillages contain chloride ions (as in saline solutions), it is important to clean up quickly. It is not possible to specify a cleaning regime which can remove all the chloride ions from a spillage on a stainless steel surface, as the corrosion takes place at an atomic level. The grade of surface finish may also affect the ease of removing the chloride ions from the surface. A stainless steel finished with a rough surface may make it harder to remove ions from within the pits and troughs than one with a smoother finish. Highly electropolished stainless steel may be easier to clean. In practice, the use of a sterile, dry, absorbent, lint-free cloth to soak up the spillage as soon as possible after it has occurred is recommended. This could be followed by cleaning down the surface with sterile water.

Some agents appear to cause discoloration. Typically, gamma-irradiated IMS 70% or IPA 70% are used as surface decontaminating agents. The irradiation process can generate aldehydes, ketones or carboxylic acids. These may be able to form complexes with the transition metals in the stainless steel. Substances added to the disinfecting agents to make them unpalatable may leave residues once the agent has evaporated. Chlorhexidine in solutions can have the same effect.

Decontamination solutions containing chlorine dioxide may not be ideal for application to stainless steel surfaces. If such solutions have to be used, corrosion effects may be minimised by immediate rinsing.

Suggested reading

1. Alexander, W and Street, A, *Metals in the Service of Man* (various editions) London, Penguin Books.
2. Higgins, R A (1993), *Engineering Metallurgy: Applied Physical Metallurgy*, 6th edn. Oxford, Butterworth-Heinemann.
3. Trethewey, K R and Chamberlain J (1995), *Corrosion for Students of Science and Engineering*. Englewood Cliffs, NJ, Prentice-Hall.
4. See also: BS 970:1983 (Part 4) and references from AISI, American Iron and Steel Institute.

Appendix 4

HEPA filtration mechanisms, MPPS and typical particle sizes

A guide to the mechanisms of filtration, how high efficiency particulate air (HEPA) filters work and the relationship to the most penetrating particle size (MPPS). Information about common particles including viruses and their relative size is provided.

A4.1 Introduction

A high efficiency particulate air (HEPA) filter is designed to remove particles of about 2 μm and smaller from an airstream. Much less expensive filters are used to remove larger particles. Long glass fibres of very small diameter are assembled into a non-woven paper-like matrix media where the fibres criss-cross randomly through the depth of the media, producing uncontrolled pore sizes, and 'random' or tortuous pathways for the passage of air. A HEPA filter comprises a long length of this media that is pleated into a compact configuration, held apart with separators and bonded into a housing.

As particles are carried through the filter media by the airstream, they collide with the fibres and with other particles that are already stuck to the fibres. When this happens, a variety of physical effects capture and retain the particle. The 'capture' effects include four principal mechanisms: straining, impaction, interception and diffusion. The 'retention' effects are very strong physical forces, e.g. electrostatic.

A4.2 Mechanisms

A4.2.1 Straining

Straining (or sieving) is simply the retention of particles larger than the pores or interstices of the filter matrix. Prefilters are normally used to reduce the number of such particles reaching the HEPA filter.

A4.2.2 Impaction

Impaction (or inertia) occurs when a particle has sufficient mass and momentum to leave the airstream as it moves around a fibre and continue in its original direction and thus collide with the fibre. For particles of smaller mass, a smaller change of direction may cause the particle to impact with the fibre.

A4.2.3 Interception

Interception occurs where the particle remains in the airstream, but the airstream is so close to the fibre that the particle touches and attaches to the fibre. Interception applies to smaller particles than impaction.

A4.2.4 Diffusion

Diffusion occurs where particles are so small that they are affected by Brownian motion. Brownian motion is when a particle changes direction and appears to demonstrate random movement as a result of collisions with molecules and other particles similarly affected in the airstream. Although a particle is of much greater mass than a molecule, the imbalance of collisions causes the random movement. This random movement greatly increases the chances of the particle encountering a fibre.

A4.3 MPPS (most penetrating particle size)

The four mechanisms of filtration as described combine in such a way that there is one size of particle at which the penetration through the filter is at its greatest. This particle size is called the MPPS or most penetrating particle size. The MPPS is typically between 0.1 and 0.2 μm for HEPA filters used in pharmaceutical applications. BS EN 1822 requires filter manufacturers to measure penetration at the MPPS. The advantage of this is that it represents the worst case. Particles both larger and smaller than the MPPS will have a lower penetration than the most penetrating particle and it is not the case that smaller particles will have a higher penetration.

It should also be noted that the penetration and efficiency are very significantly affected by the volume flow rate through a filter. The MPPS effect and the effect of flow rate are both clearly illustrated in Figure A4.1. It is evident from the figure that filter efficiency is improved by reducing the volume flow rate through a filter and is worsened by increasing the volume flow rate.

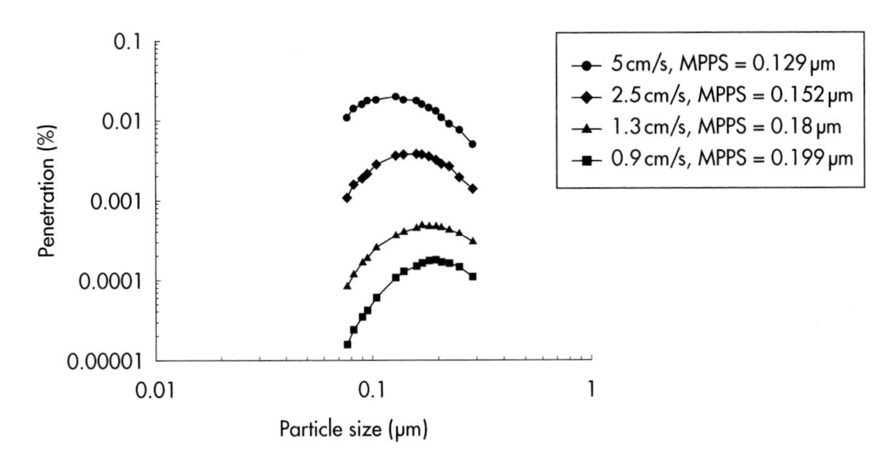

Figure A4.1 Penetration vs particle size for high permeability HEPA glass paper.

A4.3.1 Typical particle sizes

Approximately 98–99% of all particles by count are in the size range 5 µm or less. These particles tend to remain in suspension or settle out so slowly that only electrostatic and HEPA filters are effective in removing them.

A4.3.2 Common particle sizes

See Table A4.1 and Figure A4.2.

Table A4.1 Common particle sizes

Particle type	*Size (µm)*
Beach sand	100–2000
Human hair	40–300
Pollen	10–1000
Mould spores	10–30
Cement dust	3–100
Red blood cell	7.5 ± 0.3
Coal dust	1–100
Bacteria	0.3–20
Tobacco smoke	0.01–1
Oil smoke	0.03–1
Viruses	0.003–0.06
Typical atmospheric dust	0.001–30

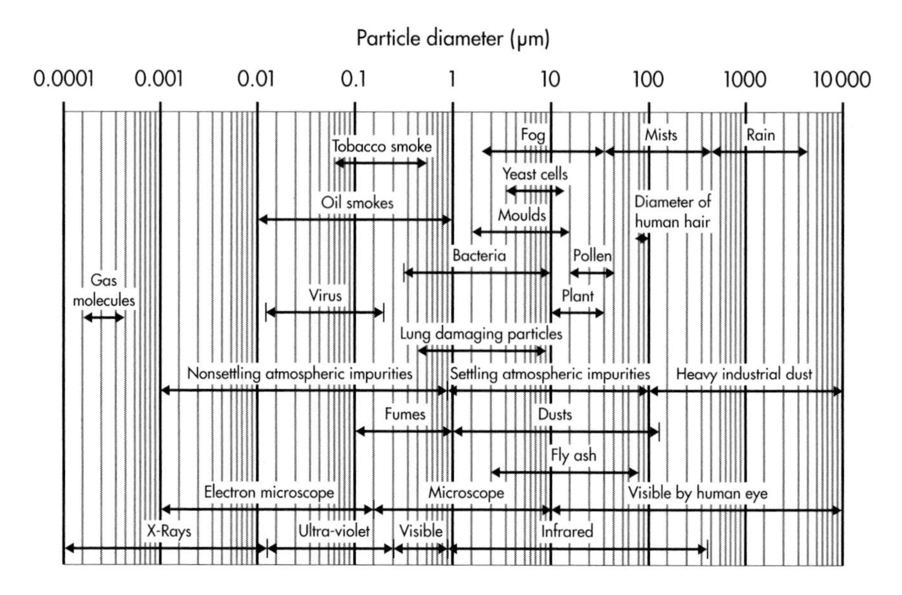

Figure A4.2 Relative size chart of common air contaminants.
The last two rows of information in the figure refer to observation or detection mechanisms and associated wavelengths of the electromagnetic spectrum for comparative purposes.

A4.4 Viruses

Concern has sometimes been expressed that virus particles, which are smaller than the particle sizes used to test a filter, will pass through a HEPA filter unhindered. This concern is unfounded for several reasons:

- Filter efficiency increases for particles that are smaller than the MPPS – see above;
- Viruses are usually attached to larger organic particles and seldom exist on their own;
- Particles with a size less than 0.3 μm are difficult to generate, requiring severe sheer forces and high energy which would disrupt a virus;
- Very small particles are seldom viable, the majority of viruses requiring a protective protein coat for viability;

Thus HEPA filters are as effective in arresting viruses as they are in arresting other, larger, particles.

Appendix 5

Calculations to estimate the size of a leak that can be detected using DOP

Supplementary information to show readers the level of sensitivity of DOP leak testing.

Scope

The formula for flow rate through a leak, as used in Chapter 8: Leak testing, may also be used to calculate the size of a leak that has been detected by a DOP test or any other challenge test such as helium.

A5.1 Calculation

The calculations show that the size of a single leak detected by the DOP test is very much smaller than the SHE of a leak quantified by a pressure decay test. This is shown in Tables A5.1 and A5.2.

The calculations are as follows:

The velocity of air through a leak (orifice) is given by the formula:

$$U = \sqrt{(2\Delta P/\rho)}$$

where:

U is velocity in m s^{-1}

ρ is density in kg m^{-3} (in dry air = 1.205 kg m^{-3} at 101.3 kPa, 20°C)

ΔP is differential pressure across the orifice in pascals

If ΔP is 100 Pa,

then U = 12.88 m s^{-1}

If V_o is volume flow rate through leak in $m^3 s^{-1}$

and A is cross-sectional area of leak in m^2,

then $V_o = U \times A$ $m^3 s^{-1}$

When the sampler head of the DOP photometer is measuring the downstream concentration, what it is detecting is the upstream challenge concentration as delivered by the volume flow rate of the leak and diluted by the much larger sampling volume flow rate.

If V_s is the sampling volume flow rate in $m^3 s^{-1}$

which is usually 1 cfm (one cubic foot per minute)

$$= 4.72 \times 10^{-4} \text{ m}^3 \text{ s}^{-1}$$

C_u is the upstream, challenge concentration

C_d is the downstream sample concentration

then $V_o \times C_u = V_s \times C_d$

and $V_o = V_s \times C_d/C_u$

Now if the DOP photometer shows a leak penetration of 0.01% (Draft ISO 14644-3),

then C_d/C_u is $0.0001 = 1 \times 10^{-4}$

and if the DOP photometer shows a penetration of 0.001% (BS 5295-1: 1989).

then C_d/C_u is $0.00001 = 1 \times 10^{-5}$

Now it is possible to calculate A in each case because

$$V_o = U \times A$$

$$= V_s \times C_d/C_u \text{ m}^3 \text{ s}^{-1}$$

so $A = V_s \times (C_d/C_u)/U$ m^2

$$= V_s \times (C_d/C_u)/U \times 10^{12} \text{ µm}^2$$

and $d = \sqrt{(4A/\pi)}$ µm

These equations, with the values above, are used to calculate the values of A and d in Table A5.1. The same equations may also be used with other challenge tests such as helium and possibly oxygen, ammonia or other gases.

A5.2 Conclusion

Tables A5.1 and A5.2 show that DOP will detect a much smaller hole than a pressure decay test.

Table A5.1 Size of hole detected by DOP test

Penetration (%)	C_d/C_u	*A = Area of SHE (μm²)*	*d = Diameter of hole (μm)*
0.01	1×10^{-4}	3664.6	68.3
0.001	1×10^{-5}	366.46	21.6

Table A5.2 SHE given by pressure decay test (for an isolator of 1 m^3) taken from Chapter 8: Leak testing

Class of isolator	*Hourly leak rate (h^{-1})*	*Diameter of SHE (μm)*
3	10^{-2}	464
2	2.5×10^{-3}	232
1	5×10^{-4}	103

Appendix 6

Activated carbon filters

Supplementary information on activated carbon filters, notes on how they are made and how they work.

A6.1 Introduction

Activated carbon is carbon that has undergone specific treatment in order to increase its adsorption characteristics. More than 70 types of activated carbons are currently marketed. Their inherent characteristics depend on the type of starting material.

Carbon filters come in several forms: powder, granules, pellets and porous cellular blocks for use in gas or liquid adsorption processes. Although the surface area of the pore structure and the adsorption capacity of activated carbons are inter-related, the total surface area is not the only factor determining the adsorption capacity of activated carbon. Hence, activated carbons with a large total surface area but with a fine microporous structure may be effective in removing odours or toxic fumes from gases but may be poor at adsorbing large amounts of coloured compounds from solutions.

The correct type of activated carbon must be selected for the particular application.

A6.2 Use

Activated carbon is often used in pharmaceutical isolators with the intention of removing hazardous materials from the isolator exhaust airstream. However, consideration must be given to the nature of the activated carbon, what disinfecting agents are being used, how the filter has been handled in transit and any other relevant factors.

The firmer the pack of the activated carbon the more resistant it is to damage during shipment and in use. Loose packs tend to allow channels to form in the filter through which some air may flow, bypassing the adsorption media. This is especially likely to happen if the filters have not been shipped or stored in their normal orientation. It is therefore good practice to place the filters in their normal orientation before fitting and shake them vigorously so the carbon granules are distributed evenly.

A further consideration is the microbiological contamination of the activated carbon filter. Activated carbon filters are hydrophilic and can support the growth of microorganisms such as bacteria, moulds and fungi. This will occur especially in humid atmospheres and as other nutrient substances are adsorbed. Therefore a HEPA filter should be fitted, as a pre-filter, to remove such microorganisms from the air before it reaches the activated carbon filter.

An activated carbon filter has a limited lifespan. Once its adsorption capacity has been reached, it needs changing. For hydrocarbon vapours, sensors are available to indicate when this is necessary. For many other vapours, proprietary test kits are available. For some less common vapours, it might be necessary to calculate or estimate the replacement interval.

Carbon filters are of interest in connection with isolators because it has been suggested that certain cytotoxic drugs can vaporise at room temperatures. At the time of writing this has still to be confirmed by independent studies. The adsorption of cytotoxic vapours, if they exist, by carbon filters, has yet to be measured as has the effect of saturation of the carbon with vapour from alcohol used for surface disinfection. As a secondary precaution, wherever practical, air passed through activated carbon filters should be exhausted at roof level and not recycled back into the room.

A6.3 Manufacture

Activated carbon filters are produced from charcoal which may be derived from carbonaceous materials such as coals, lignite, petroleum coke, peat, or natural vegetable materials such as wood and coconut shells.

The finished material is manufactured in stages:

- **Combustion.** The raw material is subjected to a combustion process in a restricted oxygen atmosphere to produce the charcoal.
- **Activation.** The charcoal is heated to around 800°C in an atmosphere of superheated steam. This breaks down the volatile substances within the micropores and drives them out. When complete, all that

is left is a carbon skeleton with many micropores. This microporous structure, known as 'activated carbon', is very efficient in the adsorption of gas molecules.

- **Impregnation**. The performance of the activated carbon may be further enhanced with the addition of specific chemical substances for specific fume removal by a process known as chemisorption.

The efficiency of the activated carbon for a group of similar substances can be greatly increased by choice of raw material, control of the activation stage and, if necessary, choice of chemical for chemisorption.

For gaseous adsorption processes the most suitable raw material is coconut shells. The high strength and fine porous structure of activated carbon derived from charcoal from coconut shells gives good adsorption and airflow. It is also more reproducible than, for example, carbon obtained from coal. Characteristics include:

- non-specific adsorption;
- large surface area;
- large internal pore volume;
- wide range of pore sizes;
- easy to modify properties;
- relatively cheap to produce.

The main disadvantage is that it is hydrophilic, and high concentrations of water vapour will impair the adsorption of other molecules.

A6.4 Characteristics of carbon filters

A6.4.1 Surface area and pore size

The pore size in the activated carbon ranges from 10 angstrom to 2 μm which covers the full range of molecule sizes. The number of pores is vast and therefore the total surface area is extremely large. The specific surface area is a measure of the surface area per unit of mass and can be between 600 $m^2 g^{-1}$ and 2000 $m^2 g^{-1}$. An amount of activated carbon as small as 10 g can have a surface area equivalent to about four football pitches. In order for the whole of this surface area to be available for adsorption, it must be kept free from particulate matter, and therefore a good pre-filter is very important.

The adsorption of gas molecules into the pores does not obstruct the airflow and therefore the pressure drop of the filter does not change throughout its life.

A6.4.2 Filtration technology

The filtration of gas molecules is based is the principle of adsorption. Adsorption may be defined as a process in which molecules of a gas, or a liquid, or a dissolved substance, or a suspended substance, condense or adhere to the surface of a solid. The greater the available surface area the greater the adsorption. There are two main mechanisms of adsorption: physical adsorption and chemisorption.

A6.4.3 Physical adsorption

Physical adsorption is non-specific. Gas molecules are carried to the proximity of the activated carbon surface by air flow and diffusion. The molecules then condense onto the surface of the pores through van der Waals' force which is an attractive force existing between atoms or molecules of all substances. Gases with high relative molecular mass, high boiling points and low heats of adsorption are better adsorbed than gases with low relative molecular mass, low boiling points and high heats of adsorption.

When a meniscus of the condensation appears at the opening of the pore, the pore is then effectively blocked for further adsorption.

A6.4.4 Chemisorption

Chemisorption happens when the adsorption capacity of activated carbon is increased by the addition of appropriate chemicals. The chemicals are chosen to react with the vapour to be adsorbed and the resulting compound is more effectively adsorbed than the original vapour. An example is impregnation by iodine compounds which react with mercury vapour to enhance adsorption. Many other chemicals can be used, and chemisorption as an approach can greatly extend the range of gases that can be removed from an airstream.

A6.4.5 Adsorptive efficiency

The adsorptive efficiency of an activated carbon filter is defined as the percentage reduction of the upstream concentration by the filter of a given substance and is primarily a function of residence time. It is also affected by temperature, humidity and age of filter.

Residence time, sometimes known as dwell time, is the time taken by the air to pass through the filter. This is the time that the air is in contact with the activated carbon.

Residence time is expressed by the formulae:

$$RT = \frac{\text{Charcoal filter bed depth (m)}}{\text{Air speed (m s}^{-1}\text{)}} \text{ or } \frac{\text{Charcoal filter bed volume (m}^3\text{)}}{\text{Air volume flow rate (m}^3\text{ s}^{-1}\text{)}}$$

Residence time is the time during which a molecule of gas can be adsorbed by the pores of the activated carbon and should be as long as possible to maximise the chances of the molecule being adsorbed. Residence time is a compromise between the depth or volume of the filter and the air speed or volume flow rate through it.

The temperature of the gas will affect the adsorptive efficiency. The higher the temperature the lower the efficiency. A high temperature can even generate desorption, with the filter releasing previously adsorbed gas. This characteristic is used industrially for solvent recovery or for regeneration of the activated carbon. Temperature should always be kept below 40°C.

Relative humidity (RH) also affects the adsorptive efficiency of the filter. Where the RH is high, molecules of water will be adsorbed leaving less available surface in the pores for gas molecules. It is therefore recommended that RH be kept below 60%.

Age can affect the adsorptive efficiency of a filter if it has not been properly stored due to 'poisoning' of the active carbon by gases or humidity in the atmosphere.

Saturation of an activated carbon filter takes place progressively as the active zone moves downstream through the filter bed. When the active zone reaches the downstream face of the filter, 'breakthrough' occurs and the filter loses its adsorptive efficiency. This is the very latest point at which the filter should be changed.

Guidelines for activated carbon filters suggest that to ensure adequate filter adsorption capacity, residence time should be at least 0.1 s and the air velocity through the filter should be less than 0.5 m s^{-1}. This translates into a bed size of 13.5 kg/1000 m^3/h airflow.

Appendix 7

Useful tables

The tables in this section appear in the relevant text of the book but are reproduced here for easy reference.

Table A7.1 Physical monitoring tests (see Table 7.1)

Test	*Type test*	*Factory test (FAT)*	*Commissioning (SAT/OQ)*[a]	*Routine maintenance test*[b] *(revalidation)*[c] *by engineer*	*User monitoring*
Gloves, sleeves, half-suits	NA	Yes	Yes	Yes	Gloves/sleeves sessionally, half-suits weekly
Installed HEPA filters	Yes	Yes	Yes	6 monthly	NA
Particle counts	Yes	Yes	Yes	3 monthly	NA
Airflow testing	Yes	Yes	Yes	3 monthly	NA
Flow visualisation and recovery	Yes	Yes	Optional	NA	NA
Pressure testing	Yes	Yes	Yes	6 monthly	Monitor continuously, record weekly
Breach	Yes	Yes	Yes	NA	NA
Alarms	Yes	Yes, calibrate set points	Yes, check set points	6 monthly, check set points	Weekly, check function
Leak testing	See Chapter 8: Leak testing, Table 8.7				

Notes
1 [a] Validation, OQ, SAT and commissioning are not the same thing. For a full explanation of these terms, refer to Chapter 10: Validation.
2 [b] Guidance on time intervals. These are generally taken from Beaney (2000; ref. 2 in Appendix 2: Training, p. 208), but note the following points:

(i) BS EN ISO 14644-2 recommends a maximum time interval of 6 months to demonstrate compliance with particle concentration limits (particle counts) for classification equal or less than ISO class 5 and suggests a maximum time interval of 24 months for what are termed optional tests. These include installed filter leakage, airflow visualisation and recovery.
(ii) For purposes of comparison there is no recommended test frequency in the current EU standard for microbiological safety cabinets. However, BS 5726-Part 4 recommends that cabinets should be regularly examined and tested by an experienced service engineer. When used in containment level 3 or level 4 laboratories (which might be considered analogous to aseptic units), cabinets may be required to be examined at least every 6 months.
(iii) A programme of testing should be established for the life time of the equipment.

3 [c] In addition revalidation should be carried out after maintenance or rectification work.

Table A7.2 Hourly leak rates (see Table 8.1)

Class	*Hourly leak rate (h^{-1})*	*Pressure integrity*	*Test methods*
1	$\leqq 5 \times 10^{-4}$	High	Oxygen, pressure change or Parjo
2	$<2.5 \times 10^{-3}$	Medium	Oxygen, pressure change or Parjo
3	$<1 \times 10^{-2}$	Low	Oxygen, pressure change or constant pressure
4	$<1 \times 10^{-1}$		Constant pressure

Table A7.3 Relating standard decay time to hourly leak rate (see Table 8.2)

Class of isolator (Draft ISO 14644-7 & ISO 10648-2)	*Hourly leak rate (h^{-1})*	*Standard decay time (min)*
3	10^{-2}	$\frac{60 \times 25}{10^{-2} \times 100\,000} = 1.5$
2	2.5×10^{-3}	$\frac{60 \times 25}{2.5 \times 10^{-3} \times 100\,000} = 6.0$
1	5×10^{-4}	$\frac{60 \times 25}{2.5 \times 10^{-4} \times 100\,000} = 30$

Table A7.4 Relating volumetric leak rate to hourly leak rate for an isolator of 1 m³ (see Table 8.3)

Class of isolator (Draft ISO 14644-7 & ISO 10648-2)	*Hourly leak rate (h^{-1})*	*Volumetric leak rate ($m^3 s^{-1}$)*
3	10^{-2}	$1 \times \frac{10^{-2}}{3600} = 2.8 \times 10^{-6}$
2	2.5×10^{-3}	$1 \times \frac{2.5 \times 10^{-3}}{3600} = 0.70 \times 10^{-6}$
1	5×10^{-4}	$1 \times \frac{5 \times 10^{-4}}{3600} = 0.14 \times 10^{-6}$

Table A7.5 Relating single hole equivalent (SHE) to hourly leak rate for an isolator of 1 m³ (see Table 8.4)

Class of isolator (ISO 14644-7 & ISO 10648-2)	*Hourly leak rate (h^{-1})*	*Leak rate ($m^3 s^{-1}$)*	*Area (m^2)*	*SHE (diameter)*	
				(m)	*(μm)*
3	10^{-2}	2.8×10^{-6}	0.17424×10^{-6}	0.464×10^{-3}	464
2	2.5×10^{-3}	0.7×10^{-6}	0.04356×10^{-6}	0.232×10^{-3}	232
1	5×10^{-4}	0.14×10^{-6}	0.0087×10^{-6}	0.103×10^{-3}	103

Table A7.6 Relating percentage volume change per hour and standard decay time to hourly leak rate (see Table 8.5)

Class of isolator	*Hourly leak rate (h^{-1})*	*Percentage volume change per hour (% h^{-1})*	*Standard decay time (for 25 Pa drop) (min)*
3	$\leqq 1 \times 10^{-2}$	$\leqq 1.0$	>1.5
2	$<2.5 \times 10^{-3}$	<0.25	>6
1	$<5 \times 10^{-4}$	<0.05	>30

Table A7.7 Relating volumetric leak rate and single hole equivalent to hourly leak rate calculated for an isolator of 1 m³ volume (see Table 8.6)

Class of isolator	*Hourly leak rate (h^{-1})*	*Volumetric leak rate ($m^3 s^{-1}$)*	*Single hole equivalent (SHE) (μm)*
3	$\leqq 1 \times 10^{-2}$	2.8×10^{-6}	464
2	$<2.5 \times 10^{-3}$	0.70×10^{-6}	232
1	$<5 \times 10^{-4}$	0.14×10^{-6}	103

Note: These values should be re-calculated for different volumes of isolator.

Table A7.8 Typical leak testing schedule (see Table 8.7)

Test	*Factory test (FAT)*	*Commissioning (SAT/OQ)*	*Routine maintenance tests (revalidation) by engineer*	*User monitoring*
Distributed leak test, e.g. DOP or helium	Yes	Yes	N/A	N/A
Leak rate measurement, e.g. pressure decay	Yes	Yes	6 monthly	Positive isolator monthly Negative isolator weekly
Leak detection	Used as distributed leak test before or during FAT and SAT		6 monthly[a]	
Gloves/sleeves, e.g. pressure decay	Yes	Yes	6 monthly	Before each working session
Half-suits, e.g. pressure decay	Yes	Yes	6 monthly	As for isolator

Note: Leak testing should also be performed after non-routine maintenance where integrity may have been compromised.

[a] Used if the measured leak rate is in excess of specification.

Table A7.9 Culture, incubation and recording microbiological monitoring (see Tables 9.2 and 9.3)

Medium	*Incubation temperature (°C)*	*Incubation time to initial visual check (h)*	*Total recommended incubation time (days)*
Tryptone soya agar (TSA)	30–35	48	7–14
Sabouraud dextrose agar (SAB)	20–25	48	7–14

Room air grade	*Suggested target levels of organisms*			
	Finger dabs[a]	*Settle plates (nominally 90 mm)*	*Surface samples*	*Active air sampling*
A (specification for the controlled workspace of isolators)	<1 per plate from 10 digits	1 per 2 plates for up to 4 h sample time[b]	<1 per contact plate or per 10 cm^2 surface sample	<1 per m^3 of air sampled
B	Not applicable	5	5	10
C	Not applicable	50	25	100
D (minimum specification for the background environment for siting isolators)	Not applicable	100	50	200

[a] Clarification of EC GMP guidelines for finger dabs which simply state <1.

[b] This represents a reduced acceptance level for settle plates compared to the EC GMP guidance.

Table A7.10 Typical frequencies to be considered for microbiological monitoring in an isolator and its background environment (see Table 9.4)

Interval	*Monitoring activity*
Sessional	• Settle plates – in transfer device and controlled workspaces • Finger dabs
Weekly	• Settle plates in transfer devices and work zone (non-operational state) • Surface sampling of transfer devices and work zone • Settle plates in isolator rooms and change rooms • Surface sampling of isolator and change rooms
Quarterly	• Active air sampling of rooms and isolator • Spraying in validations (if relevant) • Operator broth trials

Note: In addition to the above, the monitoring schedule should include control of any specific design features or identified weaknesses of a particular isolator or its workplace. Attention should be given to inanimate vectors of contamination such as intercom units, buttons on the isolator door handles and other specific design features. If the isolator is serviced by a support room which has a sink, swabs should be taken of the sink.

The above advice is weighted towards hospital aseptic production, but the principles should apply to larger manufacturing units.

Index

Page numbers in *italics* refer to figures and tables